Environmental Science and Engineering

Series Editors

Ulrich Förstner, Buchholz, Germany

Wim H. Rulkens, Department of Environmental Technology, Wageningen, The Netherlands

The ultimate goal of this series is to contribute to the protection of our environment, which calls for both profound research and the ongoing development of solutions and measurements by experts in the field. Accordingly, the series promotes not only a deeper understanding of environmental processes and the evaluation of management strategies, but also design and technology aimed at improving environmental quality. Books focusing on the former are published in the subseries Environmental Science, those focusing on the latter in the subseries Environmental Engineering.

Sukanya S. · Sabu Joseph

Environmental Radon

A Tracer for Hydrological Studies

Springer

Sukanya S.
Department of Environmental Sciences
University of Kerala
Thiruvananthapuram, Kerala, India

Sabu Joseph
Department of Environmental Sciences
University of Kerala
Thiruvananthapuram, Kerala, India

ISSN 1863-5520 ISSN 1863-5539 (electronic)
Environmental Science and Engineering
ISBN 978-981-99-2674-9 ISBN 978-981-99-2672-5 (eBook)
https://doi.org/10.1007/978-981-99-2672-5

This Springer imprint is published by the registered company Springer Nature Singapore Pte Ltd.
The registered company address is: 152 Beach Road, #21-01/04 Gateway East, Singapore 189721, Singapore

Preface

Tracing the flow of water is essential for comprehending the complex dynamics of hydrological systems. Environmental tracers could be used for solving several hydrological problems associated with flow paths, residence time, flow direction, runoff generation, hydrological connectivity, pollution source, etc. An ideal hydrological tracer is soluble, mobile, readily available, relatively conservative, and easily measurable. All these requirements are reflected in the properties of radon (^{222}Rn), an environmental radioactive isotopic tracer which is potent enough to trace many intricate sub-surface processes, especially in groundwater hydrology.

In order to acquire reliable and accurate environmental data, a thorough knowledge on analytical methodologies and selection of appropriate measurement techniques are important. However, there is currently a very diversified and discrete source of literature in the field of radon sampling and analysis. In this book, a comprehensive overview on the basics of radon sampling and analytical procedures is provided along with the possible uncertainties as well as solutions. The potential applications of radon in a wide variety of hydrological systems including connected streams, hydrothermal systems, glacial systems, lakes, floodplains, submarine groundwater systems, etc., have been reviewed and discussed in this book.

As radon tracer can be used for solving hydrological issues, the highlights of this book are useful for stakeholders to achieve UN Sustainable Development Goal-6, which addresses the sustainability of water resources. Apart from its hydrological applications, radon has been receiving worldwide attention from health sector due to its carcinogenic effects. The radiological risks posed by indoor radon in dwellings and mitigation strategies are presented.

The most relevant target audiences are hydrologists, hydrogeologists, geologists, environmental scientists, nuclear physicists, hydraulic engineers, academicians, etc. As this book also covers health implications of radon, this work will attract health physicists working on environmental radiation safety in addition to other target groups mentioned earlier. This book will be useful for postgraduate students, research scholars, postdoctoral fellows, scientists, and academicians in the field of water resources, especially for those who working in water resources management.

Thiruvananthapuram, India

Sukanya S.
Sabu Joseph

Acknowledgements

A major part of the book is the outcome of a Ph.D. thesis entitled "**Hydrochemistry of surface water-groundwater resources and characterization of hydrological processes using environmental isotopes in Karamana River Basin, Kerala, India**" by the first author, who worked as a Research Fellow in a 3-year DAE-BRNS (Government of India)-funded research project (Project No. 35/14/11/2016-BRNS/35047 dtd. 31/05/2016) executed in the University of Kerala, and thank BRNS for the financial assistance.

The fruitful discussions and interpretation of the results made by **Dr. Noble Jacob** (Scientist, BARC, Mumbai) are gratefully acknowledged. The authors thank the University of Kerala for providing all sorts of support for the research work and for the book editing work.

The authors also thank the **Permissions Granting Team (Elsevier)** for granting us permission to reproduce the material of our recently published articles in this book. The copyrights permission has been obtained for the content use (whole article along with figures and tables) of the following articles in this book:

- **Sukanya, S.**, Noble, J., Sabu Joseph (2022). **Application of radon (^{222}Rn) in hydrogeological and geological investigations: An overview**. *Chemosphere*; Volume 303, 135141. https://doi.org/10.1016/j.chemosphere.2022.135141.
- **Sukanya, S.**, Noble, J., and Sabu Joseph (2021). **Factors controlling the distribution of radon (^{222}Rn) in groundwater of a tropical mountainous river basin in southwest India.** *Chemosphere*; Volume 263, 128096. https://doi.org/10.1016/j.chemosphere.2020.128096.

The authors are thankful to the Permissions Granting Team (Elsevier) for granting us the permission to use an adapted figure (Fig. 5.1, Chap. 5) from the following article: Schroeder, E.D., (2003). Water resources. Encyclopedia of Physical Science and Technology, 3rd Edition, pp. 721–751, Elsevier (Copyrights Elsevier).

We extend our heartfelt gratitude to **Springer**, our esteemed publisher, for their invaluable support and fruitful collaboration throughout this endeavor. Special recognition is due to **Dr. Yosuke Nishida**, the Publishing Editor, whose expertise and guidance greatly contributed to the success of this project. We would also like to acknowledge the Series Editors of the Springer book series Environmental Science and Engineering, **Dr. Ulrich Förstner** and **Dr. Wim H. Rulkens**.

Furthermore, we would be remiss not to express our appreciation to the dedicated team involved in the production process. We extend our thanks to **Mr. Dharaneeswaran Sundaramurthy** (Production Coordinator), **Ms. Vidyalakshmi Velmurugan** (Project Manager), proofreaders, and copy-editors whose diligent efforts were instrumental in bringing this project to its final form.

Contents

Chapter 1
Introduction

This chapter gives a general outline of tracers used to solve various hydrological problems and discusses radon, its properties, source, and migration in the environment. Also, general applications of radon as a tracer are addressed briefly as a prelude to Chap. 2.

1.1 Tracer Hydrology

Tracers are natural or anthropogenic substances in water detected at typically very low concentrations to provide hydrological insights by following, or tracing the flow of water (Leibundgut et al. 2009; Kass 1998; Benischke 2021). Understanding the intricate dynamics of hydrological systems requires the ability to trace the movement of water (Tetzlaff et al. 2015; Cook et al. 2018). This knowledge is useful in a variety of aspects, including water quality prediction, which is often influenced by different water sources, flow paths, and transit periods, (Barnes et al. 2018; Conant et al. 2019; Frei et al. 2020) as well as understanding the effects of climate and land use/land cover changes on catchment hydrology (Tu 2009; Shrestha et al. 2018).

Questions concerning the runoff generation, flow pathways, residence time, pollution source, flow direction, etc., can all be answered using tracers (Wan et al. 2020; Minaya et al. 2021; Chang et al. 2022; Qin et al. 2022). Especially, tracer techniques are of prime importance in groundwater systems due to the limited accessibility of other observation methods for subsurface waters.

1.1.1 A Brief History of Tracer Research

The first ever attempt at a tracer experiment was conducted during the tenth century (Kass 1998), where chaff was used to trace spring sources in the Jordan basin. It

S. S. and S. Joseph, *Environmental Radon*, Environmental Science and Engineering,
https://doi.org/10.1007/978-981-99-2672-5_1

followed the slow but intriguing evolution of tracer hydrology until the mid-1960s, when tracer techniques became more widely developed, especially in groundwater recharge studies (Carlston 1964; Horton and Hawkins 1965). This breakthrough was made possible in part due to advancements in measurement techniques and the digitalization of data processing (Uhlenbrook et al. 2004). Almost about the same time, the computer era began, opening up new possibilities for environmental modeling. The growth of holistic perspectives occurred throughout this enthralling period of natural science research. In the case of tracer hydrology, this meant a greater emphasis on employing tracers to characterize hydrological system behavior, typically in conjunction with other investigative approaches.

This phase has seen the majority of the development of the core ideas of tracer hydrology. Environmental isotopes continue to play a key role in creating conceptual models and perceptual awareness of hydrological processes (Benischke 2021; Frei and Gilfedder 2021). These processes include runoff generation, runoff component separation, transit periods, recharge, and groundwater flow. The use of artificial tracers has evolved from a measuring method to a potent tool for creating, parameterizing, and validating models for solute transport in groundwater and surface water (Ward et al. 2013). Three providential turning points were among the numerous other circumstances that contributed to this evolution.

Tracer hydrology has come a long way since the early 1960s, and the formation and activity of the Association of Tracer Hydrology (ATH) have played a pivotal role in achieving several breakthroughs. With a mission to promote the use of tracer techniques, the ATH has been instrumental in driving tracer research in Europe, inspiring researchers and practitioners alike. But the impact of the ATH extends far beyond the continent, as it has also contributed to the accelerated development of isotope techniques on a global scale. The establishment of the Isotope Hydrology Laboratory at the International Atomic Energy Agency (IAEA) in Vienna in 1961 was a significant milestone that accelerated the advancement of tritium-based environmental studies. This marked the emergence of environmental tritium as a powerful tool for investigating the hydrological cycle across the world. Finally, the 20th General Assembly of the International Union of Geodesy and Geophysics in Vienna in 1991 served as a turning point, paving the way for even more innovative tracer research in the future. Through the ATH and other initiatives, the field of tracer hydrology continues to evolve and inspire new discoveries, demonstrating the value of multidisciplinary research in physics, geology, and engineering.

With the goal to bring together experimental hydrologists and modelers for a comprehensive assessment of the hydrological systems, International Commission on Tracers (ICT) was formed within the framework of the International Association of Hydrological Sciences (IAHS). As there was a considerably more raw conviction in the prospects of hydrological modeling at that time, this experimentally oriented commission faced was not received well during its embryonic phase.

However, in the following years, there was a growing integration of monitoring methods in hydrological research and applied hydrology into the international community, which confirmed this structural development. Experimental hydrology, especially catchment hydrology, is an increasingly used monitoring method for investigating hydrological processes and system functions. In particular, the calibration and validation of mathematical models are based on further hydrological research of the lines.

1.1.2 Classification of Tracers

In terms of hydrology, tracers are classified into two broad categories, viz. artificial tracers and environmental tracers (Schubert and Scholten 2021). Artificial tracers are substances intentionally introduced into the hydrological system. Artificial tracers allow labeling specific parts of a hydrological system. Commonly used applications include identification of hydrological connections, flow paths, flow directions, residence time, flow velocity estimation, hydrodynamic dispersion, infiltration, baseflow separation, etc. (Leibundgut et al. 2009). On the flip side, the application of artificial tracers is restricted spatially and temporally. For instance, this group of tracers is applied in systems with residence time preferably less than one year.

Environmental tracers are present within the natural environment generated by natural processes over long periods of time (Benischke 2021). Although the term "environmental tracer" instinctively gives one an impression that all such tracers are of natural origin, there are exceptional cases. Anthropogenic substances, viz. sulfur hexafluoride (SF_6) and cesium (Cs) classified as artificial tracers, are introduced to the hydrological cycle by means of accidents or pollution. However, the aforementioned man-made tracers are used in an analogous manner as environmental tracers (Clark and Fritz 2013).

Despite the limitations like the restricted ability to estimate contributions from single locations, environmental tracers are advantageous in a lot of ways. One of the major pros is that these tracers can be used in catchment-scale investigations. Another noteworthy advantage of environmental tracers is their capability to characterize systems with long transit times. These characteristics enable environmental tracers to investigate all components of the hydrological cycle. The major applications include investigation of precipitation processes, evapotranspiration, surface water–groundwater interaction, soil water dynamics, subsurface flow mechanisms, recharge studies, hydrograph separation, origin of water, determination of residence times (age dating), paleohydrological studies, etc. (Tetzlaff et al. 2015; Hartmann and Baker 2017; Schubert and Scholten 2021).

An ideal hydrological tracer is soluble, mobile, relatively conservative, and easily measurable. Isotope tracers are promising in this regard as they satisfy all of the above requirements. Stable environmental isotopes of ^{2}H and ^{18}O, for example, have been widely employed in tracing the hydrologic cycle, including quantifying and comprehending surface water–groundwater interactions, identifying water sources and recharge methods, and estimating the relative age of groundwater (from short-term seasonal variation to long-term age variability, viz. modern versus paleowaters).

Environmental isotopes of ^{18}O, ^{13}C, and $^{87}Sr/^{86}Sr$ play important roles in quantifying rock–water interactions, radioactive isotopes such as ^{3}H, ^{14}C, ^{36}Cl, and ^{81}Kr, among others, have been effectively employed in dating groundwater, while ^{15}N, ^{34}S, ^{37}Cl, and ^{10}B isotopes are essential tools for identifying the sources of solutes and contaminants in watershed water resources (Clark and Fritz 2013; Ješkovský et al. 2019).

Among the isotope tracers, radon is a potential investigative tool suitable for solving multiple hydrological problems. The versatility of this multifaceted tracer is explored and reviewed in the forthcoming chapter. As a prolegomenon to the next chapter, a detailed description of radon and its characteristics are addressed in Sect. 1.3.

1.2 Radon in Environment

Radon-222 (^{222}Rn) is the radioactive decay product of radium-226 (^{226}Ra), one of the products of the natural decay of uranium (^{238}U). Uranium is omnipresent in rocks and soils, and due to specific mineralogy composition, activity concentrations of uranium could be varied. Out of the 37 radioactive isotopes of radon (^{193}Rn to ^{229}Rn), ^{222}Rn has the longest half-life (3.8 days), while all others have half-lives of less than one hour. ^{222}Rn is an inert radioactive gas, colorless, odorless, and can very easily escape through the soil and into the air before its decay by alpha particle emission (Ćujić et al. 2021).

Weathering and soil-forming processes lead to the redistribution and emission of radio-elements from the native bedrocks (Sukanya et al. 2021). Radioactivity in soil, amount of radioactive ions, and γ-ray attenuation depends on its formation, migration processes during soil formation, emanation coefficient, soil cover type, thickness, porosity, bulk density, moisture, and temperature (Sahu et al. 2013; Giustini et al. 2019).

The impact of radioactivity on human health has brought the issue to the forefront of environmental concerns for both public and national authorities. The World Health Organization (WHO) and the International Atomic Energy Agency (IAEA) have highlighted radon as the primary natural source of radiation exposure for humans, with the health risks associated with inhaling or ingesting ^{222}Rn being of utmost importance. In indoor settings, geogenic radon is the primary source of significant radon activities, and research by Adepelumi et al. (2005) and Gruber et al. (2013) has further emphasized this fact. As such, identifying and mitigating the risks associated with radon exposure remains a critical issue for both public health and environmental safety (Fig. 1.1).

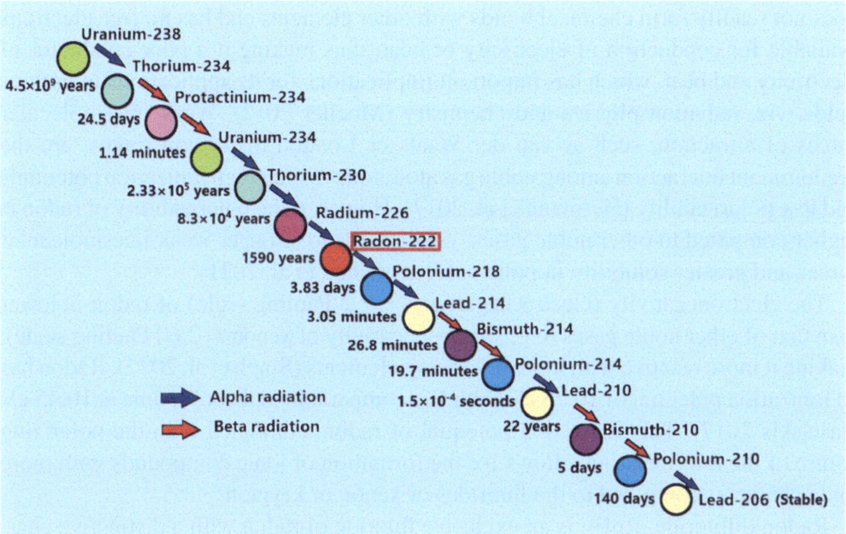

Fig. 1.1 Radon in uranium decay chain

1.2.1 Characteristics of Radon

(a) **Chemical and physical characteristics**

Radon, being a noble gas, has a closed-shell electronic structure, which renders it chemically inert due to a lack of unpaired electrons available for bonding (Quindos Poncela et al. 2013). The first ionization energy is the minimum amount of energy required to remove the outermost electron of an atom in its ground state, leading to the formation of a positively charged ion (Pfennig 2021). The first ionization energy of radon, the heaviest noble gas, is the highest among all noble gases at 1037 kJ/mol (Kang et al. 2021). This reflects the strong electrostatic attraction between the positively charged nucleus and the negatively charged electrons in its outermost shell.

In general, the first ionization energy of noble gases increases with decreasing atomic number (Kang et al. 2021). This is due to the decrease in the atomic radius, leading to a greater electrostatic attraction between the positively charged nucleus and the negatively charged electrons (Habashi et al. 2013). As a result, the energy required to remove the outermost electron becomes higher for smaller atoms with fewer electron shells. This trend is reversed for elements in other groups of the periodic table, where the first ionization energy typically decreases with increasing atomic number due to the screening effect of inner electrons (Weller et al. 2014).

The first ionization energy of radon is low, i.e., 10.7485 eV, indicating that it requires relatively little energy to remove an electron from a radon atom. However, despite its low ionization energy, radon is an inert gas with very low reactivity due to the stability of its electron configuration (Cockett et al. 2013). Because of this, radon does not readily form chemical bonds with other elements and has no free electrons available for conduction of electricity or heat, thus making it a poor conductor of electricity and heat, which has important implications for its applications in various fields, viz. radiation physics and chemistry (Moeller 2012). Weak intermolecular forces of attraction, such as van der Waals or London dispersion forces, are the predominant interaction among noble gas atoms due to their high ionization potentials and low polarizability (Hermann et al. 2017). However, the polarizability of radon is higher compared to other noble gases, which leads to stronger weak intermolecular forces and greater solubility in polar solvents (Kang et al. 2021).

The electronegativity (electro-negativity = 2.2 Pauling scale) of radon is lower than that of other noble gases (e.g., electronegativity of xenon = 2.60 Pauling scale), making it more reactive with more polarizing elements (Singh et al. 2022). Radon has an ionization potential of 10.75 eV, which is comparable to that of iodine at 10.45 eV (Jaselskis 2017). This ionization potential of radon, combined with the polarizing nature of the fluorine atom, allows for the formation of ionic compounds with more ionic character compared to the fluorides of xenon or krypton.

Radon difluoride, RnF_2, is an exclusive fluoride of radon with a distinctive characteristic of dissociating to produce ionic species, which is attributed to the ionic nature of the Rn–F bond (Fitzsimmons and Klobukowski 2013). The intermolecular forces of attraction in radon difluoride are strengthened due to its ionic character, leading to higher melting and boiling points than the difluorides of xenon or krypton. These properties also explain the low volatility of radon difluoride. The formation and properties of radon difluoride have important implications for the study of noble gas chemistry (Jaselskis 2017). Further research is needed to better understand the reactivity and properties of radon and its compounds, which could lead to the development of new technologies and applications in fields such as nuclear chemistry, environmental monitoring, and materials science.

Radon difluoride can be synthesized by heating a mixture of fluorine and radon or by increasing the concentration of radon, which produces alpha radiation that ionizes and excites the atoms, facilitating the formation of the compound (Baskaran 2016; Kang et al. 2023). However, attempts to prepare other radon compounds by various methods have been unsuccessful. The challenge of synthesizing these compounds lies in overcoming the inherent chemical inertness of radon, which can make it difficult to form bonds with other elements.

Hydrolysis of radon difluoride has not yielded an oxide or oxyfluoride, indicating that radon difluoride is the only stable fluoride of radon (Jaselskis 2017; Grandinetti 2018). Additionally, the reaction of radon with chlorine has not produced any observable reaction products.

In aqueous solutions, radon difluoride undergoes hydrolysis to produce hydrogen fluoride and oxygen. The hydrolysis reaction proceeds through the following reaction:

$$RnF_2 + H_2O \rightarrow Rn + \left(\frac{1}{2}\right)O_2 + HF \tag{1.1}$$

Radon difluoride is capable of dissolving in anhydrous hydrogen fluoride and other halogen fluorides to form ionic solutions. It has been prepared by spontaneous reaction of radon with ClF, ClF_3, ClF_5, BrF_3, BrF_5, and IF_7 at room temperature. Electromigration studies conducted in HF and HF-BrF_3 solutions have shown that radon exists as cationic and anionic species such as Rn^{2+}, RnF^+, and RnF_3^-.

The determination of thermodynamic functions for radon difluoride poses significant challenges due to the radioactivity of radon. Therefore, the Gibbs energy values can only be estimated from the reaction of radon with various oxidizing agents (Sikorska 2016). The oxidizing power of these reagents can be ranked based on the known free energies of formation for most of the halogen fluorides. The rank order of oxidizing power of these reagents is as follows:

$$ClF_5 > ClF > IF_7 > ClF_3 > BrF_5 > BrF_3 > AsF_5 > IF_5$$

Understanding the thermodynamics of radon difluoride is crucial for understanding its behavior and properties. However, the challenges associated with its radioactivity highlight the need for innovative and advanced analytical techniques to study and quantify the thermodynamic properties of this unique compound. Further studies are required to develop accurate models for the thermodynamics of radon difluoride and its reactions with other chemical species.

The potential formation of radon difluoride (RnF_2) can be inferred from the reactivity of radon with halogen fluorides, which oxidize radon except for arsenic and iodine pentafluorides. Being located in the periodic table between metals and nonmetals, radon has metalloid character and has been assigned to this group along with boron, silicon, germanium, arsenic, antimony, tellurium, polonium, and astatine (Stein 1987; Vernon 2020).

Radon can react spontaneously at 25 °C or lower temperatures with fluorine, halogen fluorides (excluding IF_5), and some oxidizing salts, although no radon compounds or ions have been observed in aqueous solutions (Stein 1986). In nonaqueous solvents such as hydrogen fluoride and halogen fluorides, cationic radon solutions have been prepared (Stein 1986; Carvalho et al. 2017).

Comparison of the properties of radon with those of krypton and xenon fluorides indicates that radon forms RnF_2 and related compounds, such as $RnF^+SbF_6^-$, $RnF^+TaF_6^-$, and $RnF^+BiF_6^-$. Reduction of RnF_2 to elemental radon by water implies that radon exists in the +2 oxidation state in RnF_2 and is reduced to elemental radon with a 0 oxidation number. The reaction can be represented as

$$RnF_2 + H_2O \rightarrow Rn + 1/2O_2 + 2\,HF \quad (1.2)$$

When Rn is heated to 400 °C with elemental fluorine, non-volatile RnF_2 is produced.

$$Rn + F_2 \rightarrow 400\,^\circ C\ RnF_2 Rn + F_2 \rightarrow 400\,^\circ C\ RnF_2 \quad (1.3)$$

Recent advances in computational methods, including first-principles calculations and crystal structure prediction, have facilitated the investigation of potential compositions of radon fluorides, despite the difficulties associated with their synthesis (Fitzsimmons and Klobukowski 2013). By combining theoretical calculations with experimental data, researchers are making significant strides in unraveling the complex chemistry of radon, which could have important implications for the development of future technologies aimed at mitigating environmental radioactivity.

In a recent study by Kang et al. (2023), the formation of radon molecules was investigated using first-principles calculations, which allowed for the prediction of possible radon fluoride compositions using a crystal structure prediction approach. The findings demonstrate the stabilization of di-, tetra-, and hexafluorides similar to that observed with xenon fluorides. In addition, coupled-cluster calculations revealed the stabilization of RnF_6 with O_h point symmetry, unlike XeF_6 which exhibits C_{3v} symmetry. These findings hold great significance in advancing our understanding of the chemical and physical properties of radon.

The parameters (boiling point, melting point, critical temperature, critical pressure, density, energy of first ionization, electronegativity, etc., of radon are summarized in Table 1.1. Though colorless in nature, radon has brilliant phosphorescence which becomes yellow at lower temperatures and orange-red at the temperature of liquid air when cooled below its freezing point. It was this property that led to radon's being called niton (i.e., the shining one) at the time of its discovery. Once formed from one of the natural radioactive series in the earth's crust, it is free to diffuse into soil, air, and the atmosphere by pressure-driven flow or further diffusion.

Recent literature on radon transport suggests that the primary mechanism for the transport of radon from soil to the atmosphere is by molecular diffusion, as supported by Schery et al. (1984) and Nazaroff (1992). The activity flux density (Jd, Bq m^{-2} s^{-1}) resulting from random molecular motion can be described by Fick's law, which states that the radon flux density is linearly proportional to its concentration gradient.

Radon diffusion coefficients in soil are primarily influenced by the soil type, pore size distribution, water content, and porosity. The molecular diffusion coefficient in air and water differ by a factor of 10^4, leading to the dominance of moisture effects in the mobility of radon in soil over other physical factors. Previous studies have demonstrated that the diffusion of radon is significantly impacted when soil

Table 1.1 Properties of radon (Weigel 1978; Lett and Adler 2013; Baskaran 2016)

Property	Value
Boiling point	−61.8 °C
Melting point	−71 °C
Critical temperature	104 °C
Critical pressure	62 atmospheres
Density at normal temperature and pressure	9.96 kg m^{-3}
Energy of first ionization	1037 kJ mol^{-1}
Electronegativity	2.3–2.5

moisture exceeds a certain threshold, which is dependent on the soil pore space geometry (Ishimori et al. 2013).

Monte Carlo analysis can be used to simulate the behavior of radon in porous media and to understand the role of different physical factors such as pore size distribution and water content on radon diffusion (Muhammad and Külahcı, 2022). Fractal theory can also be used to describe the complex structure of porous media, which can affect radon diffusion coefficients (Feng et al. 2021). The fractal dimension of porous media can be used to determine the correlation between pore sizes and shapes, which in turn can be used to estimate radon diffusion coefficients.

These tools can provide a more detailed understanding of the behavior of radon in porous media and help in predicting the transport of radon from soil to the atmosphere. Feng et al. (2019) developed a novel method for estimating the radon diffusivity in porous media by coupling fractal theory and Monte Carlo method.

(b) **Sorptive characteristics**

The sorption properties of radon are significant because radon is a radioactive gas that can accumulate in enclosed spaces, such as homes and workplaces, and poses a health risk when inhaled. The study of the sorption properties of radon can help to identify materials that can effectively adsorb or mitigate the buildup of radon in indoor environments, which is important for protecting public health (Sahu et al. 2014; Zhou et al. 2018). The sorption properties of radon can also provide information about the transport and behavior of radon in the environment, which can help to develop strategies for managing radon contamination in soil and groundwater.

Radon sorption is a complex phenomenon that is highly dependent on environmental conditions and the properties of the surface it interacts with (Guo et al. 2017; Lai et al. 2021; Maier et al. 2021). Initially, it was assumed that radon was inert and did not interact with soil or rock grain surfaces. However, later studies showed that sorption of radon to grain surfaces can occur under certain conditions (Schery and Whittlestone 1989).

The rate of sorption–desorption reactions significantly controls the mobilization of radionuclides, and the contribution from desorption is likely limited compared to other major mechanisms of release of radon to the atmospheric air over much of the Earth's surface (Saint-Fort 2018). Tanner (1980) concluded that for the dry, low-pressure environment of lunar rocks, sorption seems to be important, but the evidence for significant sorption at temperatures and humidity common at the earth's surface is conflicting. Radon is readily absorbed on charcoal, silica gel, and similar substances, a property that can be used to separate it from other gases. As a result, silica gel is not preferred as drierite material in radon monitoring devices to avoid measurement uncertainties (See Chap. 2).

The sorption of radon on various materials is a complex function of the nature of the surface and its interactions with the radon atom, including van der Waals forces, electrostatic interactions, and chemical reactions (Lai et al. 2021). The extent and rate of sorption also depend on environmental factors such as temperature, humidity, and pressure (Maier et al. 2021). Therefore, it is important to consider the complex sorption behavior of radon when studying its environmental transport and fate.

(c) **Gas–water partitioning and solubility of radon**

The equilibrium partitioning of radon gas between the gas and aqueous phases, as explained by Henry's law is expressed

$$C_w = KC_aC_w = KC_a \tag{1.4}$$

where C_a represents the radon concentration in the gas phase (Bq per m^3 of pore air), C_w is the radon concentration in water (Bq per m^3 of water), and K is the dimensional partition coefficient, which is a function of temperature.

In mathematical terms, the law is expressed as the concentration of radon in the gas phase being directly proportional to its concentration in the aqueous phase, with a proportionality constant, i.e., Henry's law constant. At 20 and 31.6 °C, the equilibrium partition coefficients of radon between air and distilled water are reported to be 0.245 and 0.193, respectively (Boyle 1911; Nazaroff 1992). This suggests that radon is more soluble in water at lower temperatures.

The partition coefficient can vary depending on the nature of the water and the temperature. In general, higher temperature and lower pH of water tend to decrease the partition coefficient, indicating greater radon release to the gas phase. Additionally, the presence of other dissolved gases or solutes in the water can affect radon partitioning. The solubility of radon in water, as represented by $K_{w/\text{air}}$, is an important factor to consider in hydrological studies as it can affect the distribution of radon between water and air phases, as well as its potential for build-up in groundwater. This relationship can be expressed as (Weigel 1978)

$$K_{w/\text{air}} = \frac{C_w}{C_{\text{air}}} \tag{1.5}$$

Eq. (1.6) represents the calculation of radon solubility in water relative to air, where C_{air} and C_w are the radon activity concentrations in air and water at equilibrium, respectively.

Ionic radon can displace ions of the first and some of the second group elements. Radon can be removed efficiently from air by using fluorinated ion exchange tubes or complex salts. The distribution coefficient of cationic radon in BrF_3^- trichlorotrifluoroethane solution ranges from ~90 to 4000 L/kg, which is related to the radon partition coefficient between water and air ($K_{w/\text{air}} = 0.25$ at 296°K). The degree of partitioning of radon is a function of its partition coefficient between water and air, which is $K_{w/\text{air}} = 0.25$ at room temperature (296°K) according to Singh et al. (2021).

The octanol–water partition coefficient of ^{222}Rn is K_{OW}:

$$K_{OW} \text{ of } {}^{222}\text{Rn} = 32.4 \pm 1.5 \text{ (at } 20^{\circ}\text{C)} \tag{1.6}$$

The gas–water partitioning of radon in saline water is an important consideration for understanding the behavior of radon in natural aquatic systems, particularly in regions with high levels of salinity, such as coastal areas or saltwater aquifers. By considering the impact of salinity on the gas–water partitioning of radon, scientists can develop a more holistic picture of radon transport and accumulation in natural saline aquatic systems, including groundwater, surface water, and coastal waters.

Schubert et al. (2012) introduced a model to depict the impact of temperature and salinity on the partitioning coefficient of radon in water, yet this model was confined to a salinity range of 0–45 ppt and a temperature threshold lower than 50 °C, with no consideration given to the influence of acidity. In order to extend the application of gas–water partitioning of radon to a broader range of diverse environments, it is imperative to develop a more comprehensive model that encompasses a wider range of variables such as salinity, temperature, and pH. Such a model would provide a more accurate representation of radon behavior and fate in aquatic environments, ultimately contributing to more effective management and protection of water resources.

1.3 Source, Migration, and Fate in Environment

The source and migration of radon in environment have to be clearly understood for developing management strategies. A million pounds (500 tons) of rocks are likely to have 1–3 pounds of uranium distributed through it (Otton 1992). As rocks are subjected to mechanical and chemical alterations to form soils at earth's surface, most soils also bear (1–3 ppm) of uranium (Sahu et al. 2014, 2016).

In general, the uranium content of soil will be comparable to the uranium content of the rock from which the soil was derived. However, some rock types, viz. light-colored volcanic rocks, dark shales, granites, phosphate-bearing sedimentary rocks, and metamorphic rocks may contain as much as 100 ppm uranium (Missimer et al. 2019; El Mezayen et al. 2022). These rock layers underlie various parts of India, the United States, Africa, China, etc. (Ray et al. 2016; Ballouard et al. 2017; Akingboye et al. 2022; Fusheng et al. 2022).

In regions with high uranium levels, it is usually anticipated that the indoor radon levels in houses will be probably high. Differing from this presupposition, some dwellings in those high uranium regions will be having low levels of indoor radon and some other buildings will be having high indoor radon levels in low uranium regions (Bezhuidenhout 2021). This is because of the fact that, in addition to the uranium content, radon activities in a dwelling are influenced by other factors. These details are documented in Chap. 8.

Each atom of radium decays by emitting an alpha particle (two neutrons and two protons) from its nucleus. When an alpha particle is released, the newly formed radon atom recoils in the opposite direction with sufficient kinetic energy ($\sim$100 keV) analogous to high-powered rifle recoil when bullet is released. This phenomenon called "alpha recoil" is the most significant factor affecting the release of radon from mineral grains (Mehta and Kocar 2018). Generally, ^{222}Rn produced via such alpha particle decay may travel approximately 20–70 nm in mineral structures (Carvalho et al. 2017; Ram et al. 2021), 77–100 nm in water and 53–63 μm in air (IAEA 2013; Sharma et al. 2020; Thu and Van Thang 2020).

When the recoil direction is towards the grain surface, the radon atom will be migrated inwards into the mineral even in cases where radium atom is located near the surface of a grain. However, few radon atoms reach the grain surface due to this recoil and leave the mineral. These atoms finally enter the pore space between grains or fractures in rocks (Sukanya et al. 2021; Przylibski et al. 2022). These newly formed radon atoms often become embedded in adjacent mineral grains after migrating through the pore space. The fraction of radon atoms produced from ^{226}Ra decay released from radium-bearing grains into the interstitial space of grains is known as the emanation coefficient (Fig. 1.2). Radium distribution, moisture content, grain size, and shape predominantly determine the emanation of radon in the soil.

Out of the total radon produced from the mineral grains, only 10–50% escapes and enters the pores (Appleton 2013; Khan et al. 2023). In the United States, most soils contain about 0.33–1 pCi of radium per gram of mineral matter and 200–2000

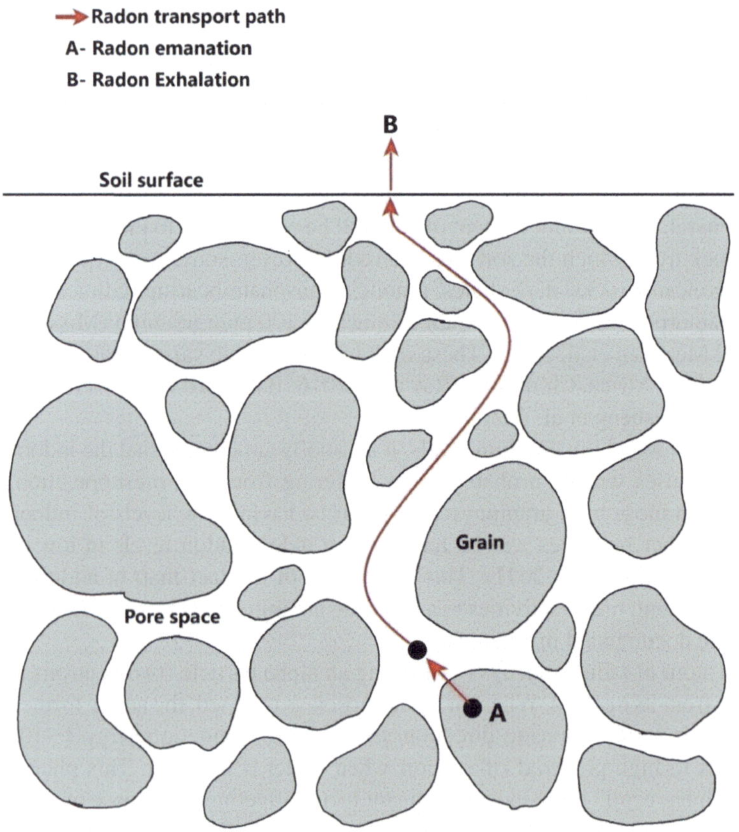

Fig. 1.2 Radon migration processes from mineral grains

pCi of radon per liter of soil air (Appleton 2013). Radon atoms entering the pore space are then transported by diffusion and advection through this space until they in turn decay or are released into the atmosphere (exhalation). Radon generation and transport in porous materials involve the solid, liquid, and gas phases in the process of emanation, diffusion, advection, absorption in the liquid phase, and adsorption in the solid phase.

Owing to its gaseous nature, the mobility of radon is much greater compared to its parent isotopes—uranium and radium adhered to solid matrix in rocks and soils. This property favors the movement of radon through fractures and openings in rocks as well as pore spaces between soil grains. The amount of water in the pore space (i.e., soil moisture content), the percentage of pore space in the soil (i.e., porosity), and the interconnectedness of the pore spaces which defines the capacity of soil to transfer water and air controls the radon movement.

When compared to impermeable soils like clays, radon can travel more easily through permeable soils like coarse sand and gravel. Radon can flow more quickly via cracks in any type of rock or soil (Feng et al. 2021; Sukanya et al. 2021). The rate of migration influences the radon concentrations inside a building (Baskaran, 2016; Ćujić et al. 2021).

Compared to radon in air, radon in water tends to move more slowly. Radon travels a very short distance in wet soils than in dry conditions before decaying (Agard and Gundersen 2020). Due to these factors, radon levels inside dwellings may be high in places with comparatively dry, permeable soils and bedrock (e.g., coarse glacial deposits, cavernous bedrock, karstic terrains, etc.).

1.3.1 Radon Entry into Buildings

Radon moving through subsurface, i.e., soil pore spaces and rock fractures escapes into the atmosphere when it reaches near earth's surface (Appleton 2013; Appleton et al. 2020). Soil air often flows towards a building foundation in three ways: (i) air pressure differences between soil and the building, (ii) the presence of openings in the foundation, and (iii) considerable permeability around the basement. Coarse gravel is usually laid down as a base for the basement slab (Angell 2013; Frutos et al. 2020). After the basement walls have been constructed, a material that is frequently more permeable than the original ground is used to fill the space between the basement walls and the ground outside. This filled gap is known as the "disturbed zone" (Appleton 2013).

From the surrounding soil, radon migrates into the disturbed zone and the gravel bed beneath. Typically, soil and rocks from the foundation site are used as backfill in the disturbed zone; these materials also release radon, known as "radon exhalation". The radon activity concentration in the disturbed zone and gravel bed is influenced by the amount of uranium in the local rock, the rock type and moisture content as well as the permeability of the soil surrounding the disturbed zone and beneath the gravel bed (Wysocka et al. 2022).

The air pressure outside most dwellings is frequently higher than the air pressure inside. As a result, air tends to enter the home through gaps in the foundation of the building from the disturbed zone and gravel bed. Cracks, utility entry, gaps between slabs and foundation walls, sumps, permeable foundation materials, and exposed soil in crawl spaces and basements are a few examples of openings that exist in all dwelling foundations (Khan et al. 2019; Saâdi et al. 2021).

On an average, most dwellings receive less than 1% of the indoor air from the soil; the rest comes from outdoor air, which typically has very low levels of radon. However, dwellings with low interior air pressures, inadequately sealed foundations, and multiple soil air entry points may receive almost 20% of the indoor air from the soil. Radon levels within the building could be extremely high, even though those levels are moderate in the soil air (Goggins et al. 2018). More details are included in Sect. 8.3.

1.3.2 Radon Migration in Atmosphere

Radon is released predominantly from the earth's surface, with approximately 99% originating from continents and 1% from oceans. The ability to quantify the source and sink of radon is facilitated by its release from the earth's surface and its removal via first-order decay rate constant, i.e., 2.1×10^{-6} s^{-1} (Ćujić et al. 2021). Radon has three isotopes, namely ^{219}Rn, ^{220}Rn, and ^{222}Rn. While ^{219}Rn is derived from the ^{235}U series, it is much less abundant than ^{222}Rn due to the relative abundance of ^{238}U and ^{235}U, with a $^{238}U/^{235}U$ atomic ratio of 137.7, and its half-life of 3.9 s, making it unsuitable for atmospheric studies.

After being released from the rocks and minerals of the earth's upper crust, ^{222}Rn undergoes radioactive decay, resulting in the production of a series of radionuclides during its atmospheric journey via diffusion and advection (Fig. 1.1). Radon activity in the planetary boundary layer (PBL) varies significantly above continental and oceanic areas (Ishijima et al. 2022). Although equilibrium between radon in surface waters and the atmosphere should preclude any exchange between these two systems, there is always nonequilibrium, and atmospheric concentrations are significantly lower than those in surface waters.

The activity of ^{222}Rn in the PBL over land is influenced by various factors, including seasonal variations in the radon source, which depend on soil moisture, pressure gradients, seasonal variations in the strength, frequency, and/or efficiency of mechanisms that lift the boundary layer air to the mid-troposphere (e.g., cold fronts, wet/dry convection, etc.), seasonal variations in the latitudinal axis of convection, and seasonal variations in the sources of air mass (Baskaran 2016; Bulko et al. 2018; Zhang et al. 2021). Changes in radon concentrations near the land surface are usually dominated by the diurnal cycle with respect to mixing depth (Victor et al. 2019; Ishijima et al. 2022). Alterations in the strength and frequency of atmospheric mixing, on the other hand, can affect the transport of radon from the PBL to the free troposphere, where it can be transported over long distances.

1.4 A Synopsis of Applications of Radon Tracer

Certain properties of radon, viz. its short half-life (3.8 days), magnitude 3–4 times higher in groundwater than surface water, preferential partitioning, sensitivity to sudden changes in subsurface conditions, ubiquity, non-invasiveness make it an ideal tracer for geological and hydrogeological investigations. Various applications of radon tracer are summarized below:

1.4.1 Atmospheric and Climate Investigations

^{222}Rn and its progeny, such as ^{210}Pb, ^{210}Bi, and ^{210}Po, have proven to be valuable atmospheric tracers in providing information on the sources of air masses and aerosols, the stability and vertical movements of air masses, the residence times and removal rate constants of aerosols, and the deposition velocities of aerosols in the planetary boundary layer (Zoran et al. 2019). The usefulness of ^{222}Rn as a tracer is due to its longer mean life of 5.51 days, also known as the e-folding life time, which is longer than typical turbulent time scales of less than 1 h. However, it is short enough to constrain ^{222}Rn concentrations in the free troposphere where the concentrations are much lower than typical atmospheric boundary layer values. Its mean life is comparable to the transit time of air masses across major continents and some oceans, but much shorter than the global mixing time scale of the atmosphere, making it widely dispersed in the atmosphere (Zhang et al. 2021).

Radon serves as an excellent atmospheric tracer for studying air circulation patterns and processes for various scales (local to global) due to its inert nature, synoptic-timescale mean lifetime, and well-characterized source function (Aba et al. 2020; Ishijima et al. 2022). The residence times of water vapor, aerosols, and other important aspects of atmospheric dynamics are comparable to the mean life of radon, making it a useful tracer. The large gradient in radon concentrations between oceanic and terrestrial radon sources is particularly helpful in identifying the sources of air masses and serving as an unambiguous indicator of recent terrestrial influence on the oceanic air mass (Röttger et al. 2021; Wu et al. 2023). Moreover, radon can be used as a tracer to evaluate dispersal models for identifying successful greenhouse gas (GHG) mitigation strategies (Röttger et al. 2021).

To increase the accuracy of both radiation protection measurements and those used for GHG modeling, traceability to SI units for radon release rates from soil, its concentration in the atmosphere, and validated models for its dispersal are required. The Radon Tracer Method (RTM) is essential for GHG emission estimates that support national reporting under the Paris Agreement on climate change. Atmospheric ^{222}Rn monitors are already being included in international and national GHG monitoring infrastructures as recommended gas. However, there is still a significant lack of data in Southern Europe and over the Mediterranean region. Additionally, harmonization of the experimental techniques used for the measurements of atmospheric ^{222}Rn activity concentrations and fluxes is necessary. Zoran et al. (2019) utilized temporal trends of ^{222}Rn to trace urban air pollution.

The radon tracer method (RTM) is a popular approach used to estimate continental nocturnal GHG fluxes on a local scale (Grossi et al. 2018). The method relies on the fact that ^{222}Rn is a naturally occurring short-lived radioactive isotope emitted from continental soils but rarely from ocean surfaces, making it a useful tracer for boundary layer mixing processes. Since its radioactive half-life is approximately 3.8 days, ^{222}Rn can accumulate in air masses dwelling over a continent, enabling it to be used as a measure of an air mass's continental contribution. The ^{222}Rn activity concentration has a significant vertical drop from elevated values in the continental boundary layer to modest activity concentrations in the free troposphere due to its short half-life.

Vertical mixing of the atmosphere plays a crucial role in determining the distribution of ^{222}Rn and its progeny in the atmosphere. When the vertical mixing is limited, such as during the nocturnal boundary layer, ^{222}Rn accumulates near the ground along with other gases that have sources near the surface. In such cases, the rate of ^{222}Rn exhalation from the Earth's surface can be estimated to calculate the flux of the gas of concern, such as CH_4. The Integrated Carbon Observation System Research Infrastructure recommends using atmospheric ^{222}Rn measurements for transport model validation and the application of the radon tracer method at ICOS atmosphere stations (Heiskanen et al. 2022). The radon tracer method has been widely used to estimate greenhouse gas emissions and sinks. However, numerous earlier studies assumed that the ^{222}Rn flux from the soil is spatially homogenous and shows very low variations based on seasonal influences.

Alegría et al. (2023) conducted a study analyzing radon measurements and meteorological parameters at two sites in the Basque country. The study investigated yearly, seasonal, and diurnal cycle differences and similarities at both sites and analyzed the temporal evolution of radon concentration during two heatwave periods. The analysis revealed two different patterns of radon concentrations under the synoptic pattern and identified regional transport channels of radon concentrations between the two sites.

^{222}Rn can be used as a tool to classify atmospheric conditions and determine the degree of dilution of pollutants (Kikaj et al. 2020, 2023). Crawford et al. (2023) analyzed 13 years of air quality data in Richmond, Australia, and found that increasing radon-derived stability categories were associated with reducing the intensity of frontal systems and smaller mixing volumes for pollutants released into the atmosphere. This shows the potential of near-surface radon observations and ventilation coefficients as effective tools for investigating air quality mitigation strategies.

1.4.2 Geological Studies

(i) **Radon as an earthquake and volcanic precursor**

Generally, the release of confined gases from the earth's crust, including ^{222}Rn, H_2, He, CO_2, CH_4, Ar, O_2, and N_2, occurs before seismic activity, which is frequently preceded by minute earth movements and stress shifts. The concentrations of these terrestrial gases may change before, during, and after an earthquake (Kamislioglu 2021; Chowdhury et al. 2022; Manisa et al. 2022; Galiana-Merino et al. 2022). Of all these gases, radon has been the most extensively studied for earthquake prediction purposes (Kamislioglu 2021; Chowdhury et al. 2022; Manisa et al. 2022; Galiana-Merino et al. 2022). This is because radon is the sole natural gaseous radioactive tracer and is not considerably influenced by chemical equilibration processes, making it a valuable tool for detecting earthquake precursor effects (Sukanya et al. 2022).

An earthquake might be anticipated by an abrupt anomaly in the measured radon activity concentrations. Since then, numerous researchers across the world have employed continuous monitoring of ^{222}Rn in the soil-gas or groundwater as a method to forecast seismic activity (Barberio et al. 2020; Chowdhury et al. 2022; Manisa et al. 2022). To investigate the aberrant radon signatures, it is advised to use a window of 30 days with the earthquake's date as the midpoint (Shukla et al. 2020). According to statistics, an anomalous signature connected to earthquakes is present when ^{222}Rn activity concentrations deviate by more than 2σ from the mean value ($m \pm 2\sigma$) (Sahoo et al. 2020; Vimercati et al. 2021; Satti et al. 2022).

Microcracks are formed in the rock mass due to regional stress before an earthquake occurs (Zoran et al. 2019; Mahmood et al. 2020). These microcracks lead to anomalous changes in radon activity concentrations, either an increase or a decline. Kuo and Tsunomori (2014) demonstrated that this radon decline can be linked to volumetric strain in aquifer rocks, and they created a mathematical model for estimating fracture porosity in a naturally fractured reservoir. It is important to understand this process since these reservoirs function as natural repositories of groundwater, geothermal, and hydrocarbon resources (Sukanya et al. 2022).

To enhance the accuracy of seismic studies, fluctuations of radon in groundwater are deemed more reliable as compared to those in soil or atmosphere that are prone to meteorological influence. In order to differentiate between seismic and non-seismic (meteorological) radon fluctuations, computational techniques are generally employed (Arora et al. 2017).

Although soil radon emissions have been employed as a tracer of volcanic activity, establishing the relationship between the two requires further investigation due to the complex radon transport process. Recently, passive dosimetry was employed to measure radon activity in diluted volcanic gases of Mt. Etna (Terray et al. 2020). This study strongly suggests for the first time that the plume is enriched with radon.

Exceptionally high radon activities up to 8 kBq/m^3 have been detected in the air in a few areas of the Mt. Etna central crater that are characterized by increased soil fracture and degassing. These findings suggest that in highly active zones, where soil radon probes cannot be installed due to high temperature and acidity, soil radon emissions could still be monitored via air measurements. However, to further investigate the potential of radon activity in the air as a tool to monitor active volcanoes, continuous measurements at high frequencies are essential.

(ii) **Radon for geochemical exploration**

The unique characteristics of radon make it an attractive tool for mineral exploration, particularly in its radioactive nature, its connection to uranium, and its inert properties, as noted by Hassan (2015) and Alsabbagh et al. (2018). Traditional mineral exploration methods often face challenges in the field, such as the time-consuming nature of the gravity method, the labor-intensive and costly seismic method, the susceptibility of the magnetic method to external magnetic disturbances, and the topographic limitations of the electromagnetic method. Radon overcomes these limitations as its detection is rapid, less costly, and less labor-intensive. Radon is a promising tool for identifying mineral deposits and assessing their potential for further exploration.

In the field of geochemical exploration, radon monitoring has been proven to be a valuable tool. As noted by Khattak et al. (2011), numerous case studies have been conducted to demonstrate the efficacy of radon detection in identifying mineralization. This is due to the fact that rocks that are rich in uranium, such as granite, phosphate, and shale, are associated with high concentrations of radon activity.

One of the reasons why radon is so useful in this regard is because it is constantly being produced in the soil as a result of the decay of uranium. This radon gas is then released into the air, allowing it to serve as a tracer for the presence of uranium mineralization. As highlighted by Kang et al. (2020), radon can be particularly effective as an ushering tool in identifying uranium deposits. Overall, radon monitoring has proven to be a valuable tool in the field of geochemical exploration due to its ability to detect the presence of uranium-rich rocks and identify potential mineralization sites.

It has been proved that this tracer has exceptional potential in identifying subsurface uranium sources due to its ability to migrate from deep underground to the surface, and its association with uranium-rich rocks (Baskaran 2016; Miklyaev and Petrova 2021). Faults provide easy escape routes for radon, making it an excellent tool for locating uranium sources. Elevated radon activity concentrations in the surface area are often indicative of the presence of subsurface uranium (Alsabbagh et al. 2018; Miklyaev et al. 2022). Radon activity concentrations at distances of around 1 m are reduced by approximately half due to its mean life of around 5 days and its diffusion coefficient in soil gas (Aydar and Diker 2021). By diffusion alone, uranium-enriched rocks and minerals, with concentrations 104 times higher than average, would be detectable at a distance of 10 m from the surface and no more than 20 m if radon moves purely by diffusion (Fleischer 1988).

The transport mechanism of radon from deep underground to the surface is a subject of ongoing research due to its importance in identifying subsurface uranium deposits. Diffusion theory suggests that radon will not migrate from distances greater than 10–15 m from the surface due to its short half-life. However, alternate transport mechanisms, such as advection through structural discontinuities, can allow for long-distance migration (>10^2 m) (Appleton 2013; Ajayi et al. 2018). Geothermally induced convection is another possible mechanism for deep underground transmission (>100 m). For an ore located at 100 m, an upward sustained gas motion speed of 3×10^{-3} cm/s is required, and the average signal velocity needed to reach the surface is >0.02 cm/s (Baskaran 2016). These transport mechanisms play a crucial role in radon monitoring and the identification of subsurface uranium deposits.

An intriguing phenomenon that has piqued the interest of geophysicists is the formation of a "halo" of anomalous radon activity concentrations corresponding to anticline structures in hydrocarbon reservoirs. This fascinating halo pattern features high radon anomalies that are concentrated in the center and low anomalies towards the edges. This can be attributed to the distribution and vertical migration of radon present in oil, gas, and water in the subsurface (Salazar et al. 2021). The implications of this phenomenon are significant, as it offers critical data regarding the geology and structure of hydrocarbon reservoirs. A thorough understanding of the distribution and concentration of radon anomalies in this context can help in locating hydrocarbon reserves with greater accuracy and efficiency, ultimately leading to more effective resource management and exploration strategies.

Fernández et al. (2016) detected abnormal radon activities in a fault system associated with a geochemically active oil reservoir. Additionally, Cvetković et al. (2021) used radon as a tracer to investigate the presence of heavier hydrocarbons and seepage in the subsurface. Nevertheless, it is advisable to integrate radon-based radiometric surveys with seismic, petrophysical, and geochemical data to prevent unsuccessful hydrocarbon projects (Salazar et al. 2021).

(iii) **Radon as a geophysical tool for locating geological structures**

Radon gas can travel long distances through faults, fractures, and other weak zones in the earth's crust, which can result in relatively high concentrations (Moreno et al. 2018). This property makes it possible to detect subsurface faults and fractures. In granitic terrains, primary porosity is generally low, but the top weathered zones can store significant amounts of water, while deeper fractures do not (Mondal 2021). However, these deeper fractures can act as effective conduits for drawing water from distant weathered zones to which they may be connected. Measuring ^{222}Rn in soil gas or groundwater is particularly useful for identifying fracture zones and exploring groundwater, especially in hard rock terrains. This approach has been demonstrated in recent research by Sukanya et al. (2021).

Radon gas has the capability to travel through faults, fractures, and weak zones within the earth's crust over long distances and can be detected in relatively high concentrations. Therefore, it is possible to locate subsurface structures such as faults and fractures by measuring radon levels (Levshenko and Grigoryan 2018). In granitic terrains, primary porosity is generally low, and deeper fractures are not capable of storing significant quantities of water. However, they can serve as good conduits for transporting water from distant weathered zones to which they may be connected. Measurement of ^{222}Rn in soil gas or groundwater is particularly useful for locating fracture zones for groundwater exploration in hard rock terrains (Sukanya et al. 2021). ^{222}Rn can also be used to identify fractures and lineaments. Gascoyne et al.

(1993) used radon gas in soil as a tracer to detect subsurface bedrock fractures and investigate local hydrogeological conditions. Reddy et al. (2006) observed diffusion and emanation as the primary mechanisms for radon release in a fractured granitic terrain in South India, rather than upward flux. Recent studies have shown the potential of ^{222}Rn in groundwater as a proxy for identifying fractures and lineaments in hard rock terrains (Sukanya et al. 2021; Arabi et al. 2018, 2021; Rengan et al. 2022). Therefore, monitoring ^{222}Rn levels in water and soil can be considered a reliable and non-invasive method for identifying active and inactive faults, lineaments, and fractures in hard rock terrains.

1.4.3 Hydrological Studies

Further, radon has wide applications in hydrological investigations including surface water–groundwater interactions, stream hydrograph separation, lake water dynamics, submarine groundwater discharge (SGD), and groundwater exploration through identifying geological faults. The applicability of radon as a tracer in hydrogeological studies is due to the fact that its activity in groundwater is about 2–4 orders of magnitude higher than surface water. Radon is particularly well suited to study groundwater–surface water interaction. Since radon readily gets escaped to atmosphere by gas transfer, its activity concentration in surface water is low.

It takes around 20 days for water entering the subsurface to attain equilibrium between radon production and decay. Radon in groundwater has been used to estimate infiltration rates in losing streams (Frei and Gilfedder 2021). To estimate residence times in the hyporheic zone, disequilibrium of radon activities in the streambed has been used (Schaper et al. 2022). Although streambed radon activities are relatively simple to measure, interpreting streambed radon profiles requires an estimation of the rate of radon production in sediments which can be very variable in heterogeneous alluvial sediments.

Similarly, some studies used radon to assess groundwater exfiltration into lakes (Petermann et al. 2018). More recently, methods for continuous radon measurement have been recommended and used in submarine groundwater discharge studies (Rocha et al. 2016; Zhang et al. 2017). This approach offers the advantage of time series, but requires time-consuming measurements in the field with the respective logistical effort. Due to changes in crustal stress/strain and variations in pore pressure, anomalies in ^{222}Rn activity concentrations have been traced along geological faults. These faults behave either as conduits or barriers with respect to groundwater flow depending on their kinematics and fault rock lithology. Hence, this application of radon is useful for groundwater exploration.

The forthcoming chapters review the state-of-the-art techniques on the measurement of radon in various domains of environment and its potential applications in various hydrological investigations, especially for water resources development and management.

References

Aba A, Ismaeel A, Al-Boloushi O, Al-Shammari H, Al-Boloushi A, Malak M (2020) Atmospheric residence times and excess of unsupported 210Po in aerosol samples from the Kuwait bay-northern gulf. Chemosphere 261:127690

Adepelumi AA, Ajayi TR, Ako BD, Ojo OA (2005) Radon soil gas as a geological mapping tool: a case study from basement complex of Nigeria. Environ Geol 48(6):762–770

Agard SS, Gundersen LC (2020) The geology and geochemistry of soils in Boyertown and Easton, Pennsylvania. In: Field studies of radon in rocks, soils, and water. CRC Press, pp 51–63

Ajayi KM, Shahbazi K, Tukkaraja P, Katzenstein K (2018) A discrete model for prediction of radon flux from fractured rocks. J Rock Mech Geotech Eng 10(5):879–892

Akingboye AS, Ademila O, Okpoli CC, Oyeshomo AV, Ijaleye RO, Faruwa AR, Adeola AO, Bery AA (2022) Radiogeochemistry, uranium migration, and radiogenic heat of the granite gneisses in parts of the southwestern basement complex of Nigeria. J Afr Earth Sc 188:104469

Alegría N, Hernández-Ceballos MÁ, Cinelli G., Peñalva I Muñoz JM (2023) Analysis of ^{222}Rn Surface Concentrations in the Basque Country (Spain): A Case Study of Heat Waves. Int J Environ Res Public Health 20(3):2105

Alsabbagh A, Alzghoul S, Marashdeh S, Abu Saleem R (2018) Examination of fault zones and uranium concentration effects on the in-soil radon levels at Central Jordan Area. J Radioanal Nucl Chem 317(1):511–517

Angell WJ (2013) Radon control in new homes: a meta-analysis of 25 years of research. Am Assoc Radon Sci Technol 45

Appleton JD, Jones DG, Miles JCH, Scivyer C (2020) Radon gas hazard. Geol Soc Lond Eng Geol Spec Publ 29(1):433–456

Appleton JD (2013) Radon in air and water. In: Essentials of medical geology. Springer, Dordrecht, pp 239–277

Arabi AS, Tukur Z, Funtua II, Ali MA, Kurowska E, Abdulhamid MS, Murtala AS (2021) Radon dynamics, activity concentration, and radiological risks associated with groundwater around tremor-prone Federal Capital Territory and environs, north-central Nigeria. Arab J Geosci 14(23):1–12

Arabi AS, Funtua II, Dewu BBM, Kwaya MY, Kurowska E, Hauwau Kulu S, Abdulhamid MS, Mahed G (2018) Geology, lineaments, and sensitivity of groundwater to radon gas contamination. Sustain Water Resour Manag 4(3):643–653

Arora BR, Kumar A, Walia V, Yang TF, Fu CC, Liu TK, Wen KL, Chen CH (2017) Assesment of the response of the meteorological/hydrological parameters on the soil gas radon emission at Hsinchu, northern Taiwan: a prerequisite to identify earthquake precursors. J Asian Earth Sci 149:49–63

Aydar E, Diker C (2021) Carcinogen soil radon enrichment in a geothermal area: case of Güzelçamlı-Davutlar district of Aydın city, western Turkey. Ecotoxicol Environ Saf 208:111466

Ballouard C, Poujol M, Boulvais P, Mercadier J, Tartèse R, Venneman T, Deloule E, Jolivet M, Kéré I, Cathelineau M, Cuney M (2017) Magmatic and hydrothermal behavior of uranium in syntectonic leucogranites: the uranium mineralization associated with the Hercynian Guérande granite (Armorican Massif, France). Ore Geol Rev 80:309–331

Barberio MD, Gori F, Barbieri M, Billi A, Casalati F, Franchini S, Lorenzetti L, Petitta M (2020) Optimization of dissolved Radon monitoring in groundwater to contribute to the evaluation of the seismic activity: an experience in central-southern Italy. SN Appl Sci 2(8):1–12

Barnes ML, Welty C, Miller AJ (2018) Impacts of development pattern on urban groundwater flow regime. Water Resour Res 54(8):5198–5212

Baskaran M (2016) Radon: A tracer for geological, geophysical and geochemical studies (Vol. 367). Basel: Springer

Benischke R (2021) Advances in the methodology and application of tracing in karst aquifers. Hydrogeol J 29(1):67–88

Bezuidenhout J (2021) Estimating indoor radon concentrations based on the uranium content of geological units in South Africa. J Environ Radioact 234:106647
Boyle RW (1911) LXXXII. The solubility of radium emanation. Application of Henry's law at low partial pressures. The London, Edinburgh, and Dublin Philosophical Magazine and Journal of Science 22(132):840–854
Bulko M, Holý K, Müllerová M (2018) On the relation between outdoor ^{222}Rn and atmospheric stability determined by a modified Turner method. J Environ Radioact 189:79–92
Carvalho F, Fernandes S, Fesenko S, Holm E, Howard B, Martin P, Phaneuf M, Porcelli D, Pröhl G, Twining J (2017) The environmental behaviour of polonium, vol 484. International Atomic Energy Agency, Vienna, Austria
Carlston CW (1964) Tritium-hydrologic research: Some results of the US Geological Survey research program. Sci 143(3608):804–806
Chowdhury S, Deb A, Barman C, Nurujjaman M, Bora DK (2022) Simultaneous monitoring of soil ^{222}Rn in the Eastern Himalayas and the geothermal region of eastern India: an earthquake precursor. Nat Hazards 112(2):1477–1502
Clark ID, Fritz P (2013) Environmental isotopes in hydrogeology. CRC Press
Cockett AH, Smith KC, Bartlett N (2013) The chemistry of the monatomic gases: pergamon texts in inorganic chemistry. Elsevier
Conant Jr B, Robinson CE, Hinton MJ, Russell HA (2019) A framework for conceptualizing groundwater-surface water interactions and identifying potential impacts on water quality, water quantity, and ecosystems. J Hydrol 574:609–627
Cook PG, Rodellas V, Stieglitz TC (2018) Quantifying surface water, porewater, and groundwater interactions using tracers: Tracer fluxes, water fluxes, and end-member concentrations. Water Resour Res 54(3):2452–2465
Crawford J, Chambers SD, Williams AG (2023) Assessing the impact of synoptic weather systems on air quality in Sydney using Radon 222. Atmos Environ 295:119537
Ćujić M, Janković Mandić L, Petrović J, Dragović R, Đorđević M, Đokić M, Dragović S (2021) Radon-222: environmental behavior and impact to (human and non-human) biota. Int J Biometeorol 65(1):69–83
Cvetković M, Kapuralić J, Pejić M, Kolenković Močilac I, Rukavina D, Smirčić D, Kamenski A, Matoš B, Špelić M (2021) Soil gas measurements of radon, CO2 and hydrocarbon concentrations as indicators of subsurface hydrocarbon accumulation and hydrocarbon seepage. Sustainability 13(7):3840
El Mezayen AM, Ibrahim EM, El-Feky MG, Omar SM, El-Shabasy AM, Taalab SA (2022) Physico-chemical conditions controlling the radionuclides mobilisation in various granitic environments. Int J Environ Anal Chem 102(4):970–986
Feng SY, Wang HQ, Cui Y, Ye YJ, Li XY, Xie D, He ZZ, Yang R (2019) Monte Carlo method for determining radon diffusion coefficients in porous media. Radiat Meas 126:106130
Feng S, Wu Y, Liu Y, Li X, Wang X, Chen P (2021) A fractal analysis of radon migration in discrete fracture network model. Chemosphere 266:129010
Fernández DP, Fusella E, Avila Y, Salas J, Teixeira D, Fernández G, Sajo-Bohus L, Greaves E, Barros H, Bolívar M, Regalado J (2016) Soil gas radon and thoron measurements in some Venezuelan oilfields. J Radioanal Nucl Chem 307(1):801–810
Fitzsimmons A, Klobukowski M (2013) Structure and stability of organic molecules containing heavy rare gas atoms. Theoret Chem Acc 132:1–10
Fleischer RL (1988) Radon in the environment—opportunities and hazards. Int J Radiat Appl Instrum Part D Nucl Tracks Radiat Meas 14(4):421–435
Frei R, Frei KM, Kristiansen SM, Jessen S, Schullehner J, Hansen B (2020) The link between surface water and groundwater-based drinking water–strontium isotope spatial distribution patterns and their relationships to Danish sediments. Appl Geochem 121:104698
Frei S, Gilfedder BS (2021) Quantifying residence times of bank filtrate: a novel framework using radon as a natural tracer. Water Res 201:117376

Frutos B, Sicilia I, Campo O, Aparicio S, González M, Anaya JJ, Rábago D, Sainz C (2020) A full-scale experimental study of sub-slab pressure fields induced by underground perforated pipes as a soil depressurisation technique in radon mitigation. J Environ Radioact 225:106420
Fusheng G, Haiyan Y, Zengqian H, Zhichun W, Ziyu L, Guocan W, Linfu X, Ye G, Wanpeng Z (2022) Structural setting of the Zoujiashan-Julong'an region, Xiangshan volcanic basin, China, interpreted from modern CSAMT data. Ore Geol Rev 105180
Galiana-Merino JJ, Molina S, Kharazian A, Toader VE, Moldovan IA, Gómez I (2022) Analysis of Radon measurements in relation to daily seismic activity rates in the Vrancea Region, Romania. Sensors 22(11):4160
Gascoyne M, Wuschke DM, Durrance EM (1993) Fracture detection and groundwater flow characterization using He and Rn in soil gases, Manitoba, Canada. Appl Geochem 8(3):223–233
Giustini F, Ciotoli G, Rinaldini A, Ruggiero L, Voltaggio M (2019) Mapping the geogenic radon potential and radon risk by using Empirical Bayesian Kriging regression: a case study from a volcanic area of central Italy. Sci Total Environ 661:449–464
Goggins J, Fuente M, Foley M (2018) Investigation of sub-slab pressure field extension in specified granular fill materials incorporating a sump-based soil depressurisation system for radon mitigation. Sci Total Environ 637:1081–1097
Grandinetti F (2018) Noble gas chemistry: structure, bonding, and gas-phase chemistry. Wiley
Grossi C, Vogel FR, Curcoll R, Àgueda A, Vargas A, Rodó X, Morguí JA (2018) Study of the daily and seasonal atmospheric CH4 mixing ratio variability in a rural Spanish region using ^{222}Rn tracer. Atmos Chem Phys 18(8):5847–5860
Gruber V, Bossew P, De Cort M, Tollefsen T (2013) The European map of the geogenic radon potential. J Radiol Prot 33:51–60
Guo L, Wang Y, Zhang L, Zeng Z, Dong W, Guo Q (2017) The temperature dependence of adsorption coefficients of 222Rn on activated charcoal: an experimental study. Appl Radiat Isot 125:185–187
Habashi F, Tsimerman V, Runge C (2013) The periodic table. Chem Educ 18:6–8
Hartmann A, Baker A (2017) Modelling karst vadose zone hydrology and its relevance for paleoclimate reconstruction. Earth Sci Rev 172:178–192
Hassan SF (2015) Depth determination of the subsurface uranium deposits using CR-39 cylindrical technique in Gneissoic granitic rocks, Abu Rushied area, Southeastern Desert, Egypt. J Radiat Res Appl Sci 8(4):562–569
Heiskanen J, Brümmer C, Buchmann N, Calfapietra C, Chen H, Gielen B, Gkritzalis T, Hammer S, Hartman S, Herbst M, Janssens IA (2022) The integrated carbon observation system in Europe. Bull Am Meteor Soc 103(3):E855–E872
Hermann J, DiStasio RA Jr, Tkatchenko A (2017) First-principles models for van der Waals interactions in molecules and materials: concepts, theory, and applications. Chem Rev 117(6):4714–4758
Horton JH, Hawkins RH (1965) Flow path of rain from the soil surface to the water table. Soil Sci 100(6):377–383
IAEA (2013) Measurement and calculation of Radon releases from NORM residues. Technical reports series No. 474. IAEA, Austria
Ishimori Y, Lange K, Martin P et al (2013) Measurement and calculation of radon releases from NORM residues. Technical Reports Series No 474.Vienna, Austria
Ishijima K, Tsuboi K, Matsueda H, Tanaka Y, Maki T, Nakamura T, Niwa Y, Hirao S (2022) Understanding temporal variations of atmospheric Radon-222 around Japan using model simulations. J Meteorol Soc Jpn Ser II 100(2):343–359
Jaselskis B (2017) Inert cases. In: Standard potentials in aqueous solution. Routledge, pp 763–785
Ješkovský M, Kaizer J, Kontuĺ I, Lujaniené G, Müllerová M, Povinec PP (2019) Analysis of environmental radionuclides. In: Handbook of radioactivity analysis, vol 2. Academic Press, pp 137–261

Kamislioglu M (2021) The use of chaotic approaches for the nonlinear analysis of soil radon gas (^{222}Rn) known as an earthquake precursor: finite ımpulse response (FIR) application. Arab J Geosci 14(7):1–6
Kang H, Chen YL, Li DP, Zhao JX, Cui FR, Xu YL (2020) Deep-penetrating geochemistry for concealed sandstone-type uranium deposits: a case study of Hadatu uranium deposit in the Erenhot Basin, North China. J Geochem Explor 211:106464
Kang J, Singh BK, Um W (2021) Efficient radon removal using fluorine-functionalized natural zeolite. J Environ Radioact 233:106607
Kang J, Park I, Shim JH, Kim DY, Um W (2023) Prediction of stable radon fluoride molecules and geometry optimization using first-principles calculations. Sci Rep 13(1):1–9
Kass W (1998) Tracing technique in geohydrology: AA Balkema. Brookfield, VT
Khan SM, Gomes J, Krewski DR (2019) Radon interventions around the globe: a systematic review. Heliyon 5(5):e01737
Khan MA, Khattak NU, Hanif M, Qadir A (2023) Assessment of soil-gas radon concentration over lithologies: a case study from district Karak, Khyber Pakhtunkhwa, Pakistan. Environ Monit Assess 195(1):1–15
Khattak NU, Khan MA, Ali N, Abbas SM (2011) Radon monitoring for geological exploration: a review. J Himal Earth Sci 44(2):91–102
Kikaj D, Chambers SD, Kobal M, Crawford J, Vaupotič J (2020) Characterizing atmospheric controls on winter urban pollution in a topographic basin setting using Radon-222. Atmos Res 237:104838
Kikaj D, Chambers SD, Crawford J, Kobal M, Gregorič A, Vaupotič J (2023) Investigating the vertical and spatial extent of radon-based classification of the atmospheric mixing state and impacts on seasonal urban air quality. Sci Total Environ 872:162126
Kuo T, Tsunomori F (2014) Estimation of fracture porosity using radon as a tracer. J Pet Sci Eng 122:700–704
Lai C, Wang Z, Qin L, Fu Y, Li B, Zhang M, Liu S, Li L, Yi H, Liu X, Zhou X (2021) Metal-organic frameworks as burgeoning materials for the capture and sensing of indoor VOCs and radon gases. Coord Chem Rev 427:213565
Leibundgut C, Maloszewski P, Külls C (2009) Tracers in hydrology. Wiley-Blackwell, Chichester, p 432
Lett JT, Adler H (eds) (2013) Advances in radiation biology, vol 9. Academic Press
Levshenko VT, Grigoryan AG (2018) Determining the positions of seismically active faults in platform regions based on the integrated profile observations. Izvestiya Phys Solid Earth 54(2):214–221
Mahmood I, Iqbal MF, Shahzad MI, Waqas A (2020) Investigation of earthquake thermal precursors in active tectonic regions of the world. J Geodyn 141:101785
Maier A, Jones J, Sternkopf S, Friedrich E, Fournier C, Kraft G (2021) Radon adsorption in charcoal. Int J Environ Res Public Health 18(9):4454
Manisa K, Erdogan M, Zedef V, Bircan H, Biçer A (2022) Variations of ^{222}Rn concentrations over active fault system in Simav, Kütahya, Western Turkey: possible causes for soil-gas ^{222}Rn anomalies. Appl Radiat Isot 190:110484
Mehta N, Kocar BD (2018) Deciphering and predicting microscale controls on radon production in soils, sediments and rock. Soil Syst 2(2):30
Miklyaev PS, Petrova TB (2021) Study of abnormal seasonal variations in the radon exhalation rate in a fault zone. Geochem Int 59(4):435–447
Miklyaev PS, Petrova TB, Shchitov DV, Sidyakin PA, Murzabekov MA, Tsebro DN, Marennyy AM, Nefedov NA, Gavriliev SG (2022) Radon transport in permeable geological environments. Sci Total Environ 852:158382
Missimer TM, Teaf C, Maliva RG, Danley-Thomson A, Covert D, Hegy M (2019) Natural radiation in the rocks, soils, and groundwater of Southern Florida with a discussion on potential health impacts. Int J Environ Res Public Health 16(10):1793
Moeller T (2012) Chemistry: with inorganic qualitative analysis. Elsevier

Mondal NC (2021) Geoelectrical signatures for detecting water-bearing zones in a micro-watershed of granitic terrain from Southern India. J Appl Geophys 191:104361
Moreno V, Bach J, Zarroca M, Font L, Roqué C, Linares R (2018) Characterization of radon levels in soil and groundwater in the North Maladeta Fault area (Central Pyrenees) and their effects on indoor radon concentration in a thermal spa. J Environ Radioact 189:1–13
Muhammad A, Külahcı F (2022) Radon transport from soil to air and Monte-Carlo simulation. J Atmos Solar Terr Phys 227:105803
Nazaroff WW (1992) Radon transport from soil to air. Rev Geophys 30(2):137–160
Otton JK (1992) The geology of radon. Government Printing Office, Washington
Petermann E, Gibson JJ, Knöller K, Pannier T, Weiß H, Schubert M (2018) Determination of groundwater discharge rates and water residence time of groundwater-fed lakes by stable isotopes of water (18O, 2H) and radon (^{222}Rn) mass balances. Hydrol Process 32(6):805–816
Pfennig BW (2021) Principles of inorganic chemistry. Wiley
Przylibski TA, Staśko S, Domin E (2022) Radon groundwater in a radon-prone area: possible uses and problems: an example from SW part of Kłodzko Valley, Sudetes, SW Poland. Environmental geochemistry and health, pp 1–17
Quindos Poncela LS, Sainz Fernandez C, Fuente Merino I, Gutierrez Villanueva JL, Gonzalez Diez A (2013) The use of radon as tracer in environmental sciences. Acta Geophys 61:848–858
Ram R, Owen ND, Pownceby MI, Duan G, Ehrig K, Etschmann B, Guagliardo P, Torpy A, Brugger J (2021) Localised solution environments drive radionuclide fractionation in uraninite. J Hazard Mater 412:125192
Ray L, Nagaraju P, Singh SP, Ravi G, Roy S (2016) Radioelemental, petrological and geochemical characterization of the Bundelkhand craton, central India: implication in the Archaean geodynamic evolution. Int J Earth Sci 105(4):1087–1107
Reddy DV, Sukhija BS, Nagabhushanam P, Reddy GK, Kumar D, Lachassagne P (2006) Soil gas radon emanometry: a tool for delineation of fractures for groundwater in granitic terrains. J Hydrol 329(1–2):186–195
Rengan AG, Joseph S, Sellamuthu S (2022) Seasonal and geological controls of radon (^{222}Rn) in groundwater of Vamanapuram river basin, SW India. Geocarto International, pp 1–31
Rocha C, Veiga-Pires C, Scholten J, Knoeller K, Gröcke DR, Carvalho L, Anibal J, Wilson J (2016) Assessing land–ocean connectivity via submarine groundwater discharge (SGD) in the Ria Formosa Lagoon (Portugal): combining radon measurements and stable isotope hydrology. Hydrol Earth Syst Sci 20(8):3077–3098
Röttger A, Röttger S, Grossi C, Vargas A, Curcoll R, Otáhal P, Hernández-Ceballos MÁ, Cinelli G, Chambers S, Barbosa SA, Ioan MR (2021) New metrology for radon at the environmental level. Meas Sci Technol 32(12):124008
Saâdi Z, Mansouri N, Marie L, Ielsch G (2021) Towards a better assessment model of transient radon concentrations in dwellings basements for the study of the effectiveness of soil radon mitigation systems designs. J Environ Radioact 235:106651
Sahoo SK, Katlamudi M, Gakka UL (2020) Singular spectrum analysis on soil radon time series (^{222}Rn) in Kachchh, Gujarat, India: detection of periodic oscillations and earthquake precursors. Arab J Geosci 13(19):1–12
Sahu P, Mishra DP, Panigrahi DC, Jha V, Patnaik RL (2013) Radon emanation from low-grade uranium ore. J Environ Radioact 126:104–114
Sahu P, Panigrahi DC, Mishra DP (2014) Sources of radon and its measurement techniques in underground uranium mines–an overview. J Sustain Mining 13(3):11–18
Sahu P, Panigrahi DC, Mishra DP (2016) A comprehensive review on sources of radon and factors affecting radon concentration in underground uranium mines. Environ Earth Sci 75(7):1–19
Salazar S, Castillo LA, Montes L (2021) Evaluation of naturally occurring radioactive materials (NORM) in the soil, in a potential area for unconventional reservoirs in the Rancheria Sub-Basin. Chemosphere 283:131098

Satti MS, Ehsan M, Abbas A, Shah M, de Oliveira-Júnior JF, Naqvi NA (2022) Atmospheric and ionospheric precursors associated with Mw $\geq$ 6.5 earthquakes from multiple satellites. J Atmos Solar Terr Phys 227:105802
Schaper JL, Zarfl C, Meinikmann K, Banks EW, Baron S, Cirpka OA, Lewandowski J (2022) Spatial variability of radon production rates in an alluvial aquifer affects travel time estimates of groundwater originating from a losing stream. Water Resour Res 58(4):e2021WR030635
Schery SD, Gaeddert DH, Wilkening MH (1984) Factors affecting exhalation of radon from a gravelly sandy loam. J Geophys Res: Atmos 89(D5):7299–7309
Schery SD, Whittlestone S (1989) Desorption of radon at the earth's surface J Geophys Res: Atmos 94(D15):18297–18303
Schubert M, Scholten J (2021) Natural radionuclides as aquatic tracers in the terrestrial and the coastal/marine environment. Water 13(5):742
Schubert M, Paschke A, Lieberman E, Burnett WC (2012) Air–water partitioning of 222Rn and its dependence on water temperature and salinity. Environ Sci Technol 46(7):3905–3911
Sharma DA, Keesari T, Rishi M, Thakur N, Pant D, Mohokar HV, Jaryal A, Kamble SN, Sinha UK (2020) Radiological and hydrological implications of dissolved radon in alluvial aquifers of western India. J Radioanal Nucl Chem 323(3):1257–1267
Shrestha S, Bhatta B, Shrestha M, Shrestha PK (2018) Integrated assessment of the climate and landuse change impact on hydrology and water quality in the Songkhram River Basin, Thailand. Sci Total Environ 643:1610–1622
Shukla V, Chauhan V, Kumar N, Hazarika D (2020) Assessment of Rn-222 continuous time series for the identification of anomalous changes during moderate earthquakes of the Garhwal Himalaya. Appl Radiat Isot 166:109327
Sikorska C (2016) Are noble gas molecules able to exhibit a superhalogen nature? RSC Adv 6(105):103418–103427
Singh P, Nautiyal OP, Joshi M, Kumar A, Ahamad T, Singh K (2021) Assessment of physicochemical and radon-attributable radiological parameters of drinking water samples of Pithoragarh district, Uttarakhand. J Radioanal Nucl Chem 330(3):1559–1570
Singh B, Garg M, Kant K (2022) Effect of different building materials on indoor radon/thoron and associated health hazards. In: Ecological and health effects of building materials. Springer, Cham, pp 467–487
Stein L (1987) Ion-exchange reactions of cationic radon. J Fluorine Chem 35(1):100
Stein L (1986) Chemical properties of radon (No. CONF-860425--27). Argonne National Lab
Sukanya S, Noble J, Joseph S (2021) Factors controlling the distribution of radon (222Rn) in groundwater of a tropical mountainous river basin in southwest India. Chemosphere 263:128096
Sukanya S, Noble J, Joseph S (2022) Application of radon (222Rn) as an environmental tracer in hydrogeological and geological investigations: an overview. Chemosphere 135141
Terray L, Gauthier PJ, Breton V, Giammanco S, Sigmarsson O, Salerno G, Caltabiano T, Falvard A (2020) Radon activity in volcanic gases of Mt. Etna by passive dosimetry. J Geophys Res Solid Earth 125(9):e2019JB019149
Tetzlaff D, Buttle J, Carey SK, McGuire K, Laudon H, Soulsby C (2015) Tracer-based assessment of flow paths, storage and runoff generation in northern catchments: a review. Hydrol Process 29(16):3475–3490
Thu HNP, Van Thang N (2020) The effects of some soil characteristics on radon emanation and diffusion. J Environ Radioact 216:106189
Tu J (2009) Combined impact of climate and land use changes on streamflow and water quality in eastern Massachusetts, USA. J Hydrol 379(3–4):268–283
Uhlenbrook S, Roser S, Tilch N (2004) Hydrological process representation at the meso-scale: the potential of a distributed, conceptual catchment model. J Hydrol 291(3–4):278–296
Vernon RE (2020) Organising the metals and nonmetals. Found Chem 22(2):217–233
Victor NJ, Siingh D, Singh RP, Singh R, Kamra AK (2019) Diurnal and seasonal variations of radon (222Rn) and their dependence on soil moisture and vertical stability of the lower atmosphere at Pune, India. J Atmos Solar Terr Phys 195:105118

Vimercati L, Cavone D, Delfino MC, De Maria L, Caputi A, Sponselli S, Corrado V, Bruno V, Spalluto G, Eranio G, Ferri GM (2021) Relationships among Indoor Radon, earthquake magnitude data and lung cancer risks in a residential building of an Apulian Town (Southern Italy). Atmosphere 12(10):1342

Ward AS, Payn RA, Gooseff MN, McGlynn BL, Bencala KE, Kelleher CA, Wondzell SM, Wagener T (2013) Variations in surface water-ground water interactions along a headwater mountain stream: comparisons between transient storage and water balance analyses. Water Resour Res 49(6):3359–3374

Weigel F (1978) Radon. Chem Ztg 102(9):287–299

Weller M, Weller MT, Overton T, Rourke J, Armstrong F (2014) Inorganic chemistry. Oxford University Press, USA

Wu C, Trounce H, Dunne E, Griffith DW, Chambers SD, Williams AG, Humphries RS, Cravigan LT, Miljevic B, Zhang C, Wang H (2023) Atmospheric concentrations and sources of black carbon over tropical Australian waters. Sci Total Environ 856:159143

Wysocka M, Nowak S, Chałupnik S, Bonczyk M (2022) Radon concentrations in dwellings in the mining area—are there observed effects of the coal mine closure? Int J Environ Res Public Health 19(9):5214

Zhang Y, Li H, Xiao K, Wang X, Lu X, Zhang M, An A, Qu W, Wan L, Zheng C, Wang X (2017) Improving estimation of submarine groundwater discharge using radium and radon tracers: application in Jiaozhou Bay, China. J Geophys Res Oceans 122(10):8263–8277

Zhang B, Liu H, Crawford JH, Chen G, Fairlie TD, Chambers S, Kang CH, Williams AG, Zhang K, Considine DB, Sulprizio MP (2021) Simulation of radon-222 with the GEOS-Chem global model: emissions, seasonality, and convective transport. Atmos Chem Phys 21(3):1861–1887

Zhou Q, Zhao G, Xiao D, Qiu S, Lei Q, Kearfott KJ (2018) Prediction of radon removal efficiency for a flow-through activated charcoal system and radon mitigation characteristics. Radiat Meas 119:112–120

Zoran MA, Savastru RS, Savastru DM, Penache MCV (2019) Temporal trends of carbon monoxide (CO) and radon (^{222}Rn) tracers of urban air pollution. J Radioanal Nucl Chem 320(1):55–70

Chapter 2
Radon Measurement Techniques

2.1 Introduction

The primary objective of this chapter is to provide an overview of well-established, standard analytical techniques used for radon detection, measurement, and monitoring. Numerous analytical methods are those that have been approved by internationally recognized groups, federal agencies, and organizations, viz. Environmental Protection Agency (EPA), American Public Health (APHA), American Association of Radon Scientists and Technologists (AARST), National Radon Proficiency Program (NRPP), etc. (Warkentin et al. 2020; Saha et al. 2021; Gopalakrishnan and Jeyanthi 2022). Moreover, important aspects like accuracy, precision, and uncertainties associated with these techniques have also been described. Although there are some approaches based on the detection of emitted gamma rays, the majority of the techniques for measuring radon and its progenies in environmental samples (air, soil, and water) rely on the detection of alpha particles released as a result of radioactive decay.

EPA issued updates regarding radon measurement techniques in 1992 and provided general guidelines for optimal measurement conditions, device placement, and documentation of results (EPA 1992). Technical guidelines for radon concentration measurements in dwellings have also been released by EPA.

2.2 Radon Analytical Methods

Two fundamental types of analytical processes are typically used to measure radon activity concentrations: (a) passive mode, in which radon enters the instrument passively through diffusion; and (b) active mode, in which radon is drawn into a radon detecting device (Janik et al. 2014; Bem et al. 2020). For radon in water, two sampling techniques are used: instantaneous or grab sampling (single point samples obtained over a short period of time), and continuous/active approach (time series

S. S. and S. Joseph, *Environmental Radon*, Environmental Science and Engineering,
https://doi.org/10.1007/978-981-99-2672-5_2

activities). Counting duration ranges from seconds to minutes, whereas monitoring time might last for several weeks, months, or even years, depending on the study objective.

2.3 Radon Measurement Methods

Radon in air, water, and soil environment can be measured by many methods. The most commonly used radon detection methods are broadly grouped into six categories and are described below.

2.3.1 Scintillation or Lucas Cell

The measurement of radon in air and water can be effectively performed using an alpha scintillation cell, commonly referred to as a Lucas cell. This method involves capturing the radon gas or emanated radon from water in the scintillation cell, which then filters out the daughter isotopes through a specialized 0.01-micron filter (Sukanya et al. 2022). The scintillation cell is coated with a layer of silver-activated ZnS, which produces scintillation upon alpha particle contact resulting from radon decay (Engelbrecht 2020).

The scintillation photons are counted using a photomultiplier tube placed at the top of the cell. The data is transmitted to a digital data logger for recording and analysis. The counts are typically expressed in Bq/m^3 (disintegrations per second per m^3) with a 2σ uncertainty level. The implementation of a specialized high-efficiency ZnS (Ag) coating in the Lucas cell has led to the development of an exceptionally sensitive detector that exhibits remarkable stability and remains unaffected by fluctuations in temperature, humidity, or dust accumulation (Abdalla and Al-Hajry 2015). With a volumetric capacity of around 272 mL and an alpha energy range spanning from 4.5 to 9 meV, the Lucas cell represents an instrument capable of delivering precise and accurate measurements over extended periods of time (Abdalla and Al-Hajry 2015; Abdalla et al. 2017).

This method has been well documented in previous studies (Sethy et al. 2014; Abdalla et al. 2021, 2022) and is a reliable approach for radon measurement in air and water. In a recent research conducted by Abdalla et al. (2022), the efficacy and sensitivity of an active radon cell coated with iodine-activated ZnS were thoroughly examined. The nanostructured ZnS:I coating material exhibited superior scintillation decay and resolution compared to the reference ZnS:Ag sample, resulting in improved detection accuracy. The scintillation response was measured under pulsed X-ray, revealing a rapid decay time (=1.492 μs), indicating the potential of the developed material for radon detection applications.

2.3.2 Semiconductor Detector

The utilization of semiconductor detectors provides an effective means of measuring radon in both air and water. In semiconductor detectors, the fundamental information carriers are electron–hole pairs, which aid in signal detection (Venkataraman 2020). Incident nuclear particles on the depletion layer of a reverse-biased p–n junction generate an electric pulse across a resistance in the external circuit, which facilitates detection and counting.

Silicon, germanium, and other crystalline matrices are used in this detector to detect charged particles or gamma rays (L'Annunziata 2020). The radon activity is measured in a silicon semiconductor detector by counting the alpha activity of polonium daughters. This instrument exhibits high stability in counting and is ideal for determining very low radon activity in the natural environment (Tokonami et al. 1996).

Irlinger et al. (2016) developed a new electronic radon/thoron monitor with semiconductor detectors, capable of capturing alpha particle energy spectra. The alpha particle spectra were simulated based on Monte Carlo application using the Geant4 toolkit, achieving reasonable agreement between measured and simulated spectra for both radon isotopes. The simulation accurately predicted calibration factors and alpha particle energy spectra, facilitating the interpretation of spectra measured in mixed radon/thoron atmospheres.

Silicon-based detectors offer superior energy resolution and linearity while maintaining rapid response time, compact size, and low susceptibility to background gamma radiation and neutron flux (Syuryavin et al. 2020; Dimitrova et al. 2023). This makes them well-suited for radon detection (Dearnaley and Whitehead 1961; Laubenstein and Lawson 2020; Venkataraman 2020).

2.3.3 Ionization Chamber

This technique is appropriate for measuring radon levels in air and water. The ionization chamber method works by generating electron–ion pairs via the ionization process when a charged particle passes through a gas or liquid (Álvarez et al. 2013). In this technique, an electret applies an electric field across two or more electrodes within a gas volume inside an ionization chamber (Steinhauser and Buchtela 2020).

Radon gas enters the chamber through a filtered inlet by passive diffusion and causes ionization as it decays. Consequently, ions generated by radon decay products are collected by the charged electret (Sukanya et al. 2022). The electret's charge decrease is proportional to the integrated radon activity over the exposure period. The total ionization in the chamber can be measured, or the pulses caused by individual alpha particles can be counted separately (Seo and Kim 2021). The advantage of the latter method is that radon and decay product pulses can be distinguished from one another.

Some of the advantages of ionization chamber are precise measurement, high sensitivity, ruggedness, and simplicity of operation, among others. One disadvantage is that it is sensitive to background gamma radiations, and therefore, corrections accounting for the background radiation have to be applied (Tracy 2010). A classic example is the Geiger–Müller tube. Recently, there have been attempts from the scientific community to develop low-cost, battery-operated ionization chambers. Studnička et al. (2019) developed a simple air-ionization, low-cost radon detector (minimum detectable activity = ~50 Bq/m^3) incorporated with an Internet of Things (IoT) sensor grid. However, this sensor was not designed to distinguish different isotopes and was found to be suitable for monitoring purposes in mines and earthquake-prone regions.

2.3.4 Liquid Scintillation Counter (LSC)

The technique of liquid scintillation counting (LSC) is well suited for measuring radon in water. This method exploits the high solubility of gaseous radon in aromatic solvents such as toluene, which was first introduced by Horrocks and Studier in 1964. Water samples are mixed directly with scintillation cocktails to form a two-phase aqueous/organic system, in which the decay products of ^{222}Rn remain in the water phase while ^{222}Rn is extracted into the organic phase (Bem et al. 2017, 2021). The sample is then stored for a build-up time of three hours until equilibrium is reached between ^{222}Rn and its alpha-emitting decay products.

The alpha activity of radon and its decay products are measured by detecting photons emitted from the scintillation fluid using a liquid scintillation counter (LSC) (Salonen et al. 2012). To reduce errors due to the interference of photo- and chemiluminescence, the sample is stored in a dark place for a few hours before measurement (Jobbágy et al. 2019).

The detection limit is a crucial metric for evaluating the performance of a radon measurement method, as it determines the lowest concentration of radon that can be reliably detected. To achieve lower detection limits, there are several considerations to be made, including the choice of vial material and alpha/beta discrimination. Vial material selection is critical, with glass and plastic being common options. Glass vials generally offer superior performance due to their lower intrinsic background radiation. On the other hand, plastic vials may be more susceptible to radon diffusion, leading to increased measurement uncertainty (Jobbágy et al. 2019).

Another approach to improve detection limits is alpha/beta discrimination, which distinguishes between alpha and beta particles (Hou 2018; Ryu et al. 2022). Alpha particles are more indicative of radon and its decay products, and separating them from beta particles can reduce the overall background signal, leading to improved detection limits. Implementing alpha/beta discrimination requires additional time and materials, which may not always be practical. However, in 2016, Stojković et al. attempted the assessment and optimization of a liquid scintillation counting (LSC) procedure for determining ^{222}Rn in water. A highly sensitive ultra-low background spectrometer equipped with Pulse Shape Analysis (PSA) circuit, which distinguishes alpha/beta spectra, was utilized for the measurements.

Notably, the calibration procedure relied on a ^{226}Ra standard having both alpha and beta progenies, emphasizing the critical importance of the PSA discriminator in providing accurate spectra separation. The calibration was improved by varying the PSA discriminator level and ^{226}Ra standard activity and assessing quench effects through a quench calibration curve. The modified procedure was evaluated using prepared ^{226}Ra solution samples and water samples with measurement uncertainty considered.

Overall, selecting the appropriate vial material and employing alpha/beta discrimination are two potential strategies to achieve a lower detection limit which is in accordance with ISO 11929:2010. Ultimately, the right approach depends on the specific measurement needs and balancing performance with practical considerations.

LSC requires a small volume of sample, typically around 6 mL, and a short measurement interval of 600 s (Celaya et al. 2018). Recently, Pujol et al. (2022) validated a method for measuring ^{222}Rn in non-saline water samples using LSC based on ISO/IEC 17,025 criteria. They found that the relative uncertainties were dependent on sample activity, ranging from 11.2% for 2 Bq/L to 4.1% for 200 Bq/L (coverage factor $k = 1$).

While LSC is a reliable and efficient method, in situ measurement is not possible. In this case, Higuchi et al. (2019) compared LSC with the RAD7 semiconductor detector degassing method and found that the RAD7 device can be used as an alternative when the conditions are not favorable for conducting LSC.

2.3.5 *Gamma-Ray Spectrometry*

This method for radon measurement in water and soil utilizes gamma-ray spectrometry to count the gamma rays emitted by the short-lived ^{222}Rn daughters, ^{214}Bi and ^{214}Pb, during their decay process (ISO 2019). A secular equilibrium between radon and its daughters is generally established within 3 h, which allows for accurate measurement of radon activity (Bonczyk and Samolej 2019; Dimitrova et al. 2023). Standard gamma-ray spectrometers with HPGe (Li) or NaI (Tl) detectors can be used for this purpose. In cases where ^{226}Ra is present in the water sample, a second measurement is required to account for ^{222}Rn in-growth from ^{226}Ra once secular equilibrium is established, which can complicate the measurement process (Jobaggy et al. 2017).

In gamma-ray spectrometry, background sources include the detector, spectrometer shield, and ambient radiation (Bučar et al. 2012). The sample matrix attenuates background radiation, except for detector contamination, with attenuation factors dependent on counting geometry and sample material (Korun et al. 2014, 2016). Shielding factors indicate the influence of a sample on spectrometer background, i.e., the ratio of background count rate with and without the sample. Determining shielding factors for each peak in the background spectrum is necessary for spectral analysis since background count rates are subtracted from peak count rates.

In 2014, Korun et al. proposed a simple model which accounted for the dimensions of the sample and its mass attenuation coefficient. Shielding factors were determined for six spectrometers based on the contributions of the detector, spectrometer shield, and ambient radiation to the spectrometer background. The determined shielding factors were in the range of 0.88–1.00 for a spectrometer with radon diffusion and 0.95–1.00 for a water sample with nuclides from the thorium decay series. Over a 25-year period, Korun et al. (2023) measured the count rate in the 2223 keV gamma-ray spectrum peak resulting from neutron capture on hydrogen in the spectrometer's shield. They found a correlation between the count rate and solar activity through the interaction between the solar wind and cosmic rays. The correlation showed both long-term and short-term patterns linked to the 11-year solar cycle and sunspots, respectively.

Further, this method is disadvantageous as there is a high possibility for radon escape from measurement container, i.e., sample container tightness is very important in determining the reliability of results (Bonczyk and Samolej 2019). Mertes et al. (2020) proposed a new emanation source for ^{222}Rn by electrodeposition of ^{226}Ra on stainless-steel discs with uncertainties less than 2% ($k = 1$). It was concluded that emission probability and detection efficiency are not required to be known when sealed ^{222}Rn tight sources of same type and geometry are used for gamma-ray spectrometric measurements.

2.3.6 Solid-State Nuclear Track Detector (SSNTD)

This technique of utilizing solid-state nuclear track detectors (SSNTD) or Etched Track Detectors (ETD) to measure radon levels in the air has been demonstrated as highly effective for mineral exploration purposes, even in challenging environments. The path taken by high-energy, nuclear particles through most insulating solids generates narrow channels of concentrated damage at an atomic level (Durrani and Bull 2013). These latent pathways can be discerned from the intrinsic distortions and disruptions to the crystal structure. Furthermore, these tracks can be exposed and rendered visible using a suitable chemical reagent when examined under a conventional optical microscope (Banjanac et al. 2006; Joshirao et al. 2013). It is important to note that, while this technique is appropriate for assessing radon levels in air, it is not suitable for determining radon activity in water.

2.4 Radon Measurement in Air

Devices for measuring air radon and its progeny can be broadly divided into two categories: passive devices and continuous monitor devices (Sukanya et al. 2022). Air is allowed to diffuse into a sensor chamber via passive radon monitors, which operate without the requirement of a power source. However, passive monitors generally provide average concentrations for the entire sampling period (typically lasting at least 48 h), and radon concentrations are often determined using the laboratory setup (Tchorz-Trzeciakiewicz et al. 2020).

Radon gas is measured by continuous radon monitors (CRMs), while radon progeny is assessed by continuous working level monitors (CWMs). Although the continuous monitoring devices need a power supply, these instruments can record and monitor radon concentrations in time intervals of less than an hour (Di Carlo et al. 2022). CRMs are commercially available to radon testing experts or building inspectors. The following is a description of how radon detectors function.

The subsequent paragraphs give an account of the basic principles on which each type of radon detector operates.

For detecting radon in air samples, low-cost passive activated charcoal adsorption detectors are available (Blanco-Novoa et al. 2018). A typical detector consists of a cylindrical, 6–10 cm in diameter, 2.5 cm deep container that is filled with 25–100 g of activated charcoal. The container has a screen attached to one side that covers the charcoal sample and allows air to diffuse in. Since these detectors are passive, radon can continuously be absorbed and de-absorbed, and during the measurement period, the adsorbed radon goes through radioactive decay (Maier et al. 2021). After a brief exposure time of about 2–7 days, the charcoal detectors are brought back to a lab where direct analysis is done by counting the gamma rays released by radon decay products on the charcoal using a sodium iodide gamma detector (Gubanski et al. 2019).

The detector can be used with either a single-channel analyzer or a multi-channel gamma spectrometer, as long as the window is configured to include the proper gamma energy window. To determine the calibration factors for the device, the detector system and detector geometry must comply (EPA 1992). As an alternative, the sample could be desorbed using an aromatic solvent (usually toluene or benzene) and then analyzed using liquid scintillation counting with a suitable fluor solution.

Alpha track detection systems are also widely used to test indoor radon levels (EPA 1992). The detector is made of a tiny piece of plastic or film that is encased in a container with a filter-covered opening or another similar design that allows radon in but prevents its progeny from entering. The cellulose nitrate film (LR-115), the thermoset polymer plastic (CR-39), and the polycarbonate plastic (Makrofol) are a few common materials used in this capacity for radon detection (Cahyadi et al. 2019; Pérez et al. 2021; El Ghazaly et al. 2021). Radon gas diffuses into the container, and the radon and its progeny emit alpha particles that hit the detector and inflict minuscule damage to the plastic material inside (do Nascimento Santos et al. 2021).

The plastic detectors are immersed in a caustic solution after the analysis to highlight these submicroscopic damage tracks and count them under a microscope or with an automated counting system (Frutos-Puerto et al. 2021). Using a conversion factor established from data generated at a laboratory, the number of tracks per unit area is connected to the radon concentration in the air. The ratio of the radon concentration to the number of tracks produced per unit of analyzed detector area per unit of time (after subtracting the background) (Van Dung et al. 2022). Alpha track detectors are advantageous over charcoal adsorption detectors, as they can be utilized for prolonged measurements (90 days–1 year), implying this method as an accurate time-integrated option. Commercial alpha track detection kits are available at affordable rates for public use.

Radon detectors using electret ion chambers (EC) capture the ions produced in the chamber by radiation from radon and its progeny using an electrostatically charged disk (EPA 1992). Through filtered apertures, radon diffuses into the chamber. As radon decays, ions are continuously produced. These ions and their progeny are drawn to the surface of the electret, which lowers the surface voltage. Given the duration of exposure period, the voltage change in the electrostatic voltmeter can be used to calculate the average radon concentration (Sorimachi et al. 2021).

For field measurements and occupational settings, flow-through alpha scintillation cells are often used to assess radon concentrations in air (NCRP 1988; Abdalla et al. 2022). The cell is made of a silver-activated zinc sulfide (ZnS) phosphor screen that, when bombarded by alpha particles, emits visible light photons (Lucas, 1957; Abdalla et al. 2021). An air pump constantly draws air through the cell, which is connected to a photomultiplier tube for continuous analysis. The photomultiplier tube records the scintillations or light flashes produced by the alpha particles from radon and its progeny that hit the ZnS screen. Radon gas concentration may be determined from the rate at which the pulses are recorded using suitable calibration and decay scheme factors (Abdalla et al. 2019).

Personal dosimeters are frequently used to evaluate radon exposure in the home and workplace. In 1967, Geiger described an early workplace dosimeter, the radon film badge, which consisted of a plastic holder containing a nuclear track film that could detect alpha particles emitted into the air (Geiger 1967). Radon gas diffused through the central opening of the badge and into the film emulsion, where the number of alpha particles was determined by counting the tracks in the processed film.

Another passive radon dosimeter utilizes a polycarbonate detector and a porous fiberglass filter to collect the radon progeny, ^{218}Po and ^{214}Po. To attenuate the energy of emitted alpha particles, a thin aluminum foil is placed between the filter and the detector. For surface trap methods, the activity is measured at the surface of objects, such as glass, that are present in the location of interest during the exposure assessment period (Malinovsky et al. 2019).

2.5 Radon Measurement in Soil

The measurement of radon in soil is accomplished using a silicon detector that records the alpha particle emissions of radon present in a measurement chamber optimized for the purpose. The chamber has a diameter of 60 mm, a length of 570 mm, and a depth of 0.7–0.8 m. The detector measures alpha emissions generated during pre-established time intervals ranging from 15 to 240 min, and a microprocessor is used to store the measured values (Papastefanou 2002). Readout can be performed by an on-site reader or a PC-compatible computer with software. The detector is a light-tight implanted silicon junction with an aluminum layer for light protection and a cellulose varnish for mechanical protection (Antonopoulos-Domis et al. 2009). The useful area of detector is 450 mm^2, and its depleted depth is 100 μm (Galli et al. 2019).

The detector resolution in air is 60 keV (FWHM) at 5.486 meV (241Am), and its sensitivity is 0.02 pulses per hour for 1 Bq m^{-3} (Papastefanou 2007; Vajda et al. 2020). The detector's saturation volumetric activity is 3 MBq m^{-3}, and its background count rate is below 1 event every 24 h. The detection limit for radon is 50 Bq m^{-3}. The detector also measures atmospheric parameters such as temperature (with a resolution of 0.1 °C for variation), atmospheric pressure (with a resolution of 1 hPa), and rain precipitation (with a resolution of 0.1 mm). It can also allow real-time tracking of radon measurements worldwide using modem or satellite modules, with a 1-year autonomy provided by two D-type alkaline batteries.

Barasol (ALGADE, Route Nationale 2, F87250 Bessines-sur-Gartempe, France) detectors are used for static measurements of radon gas without causing any disturbance. Another example of a radon measurement system is SARAD RTM 2200, which is capable of quantifying the concentration of radon (^{222}Rn) and thoron (^{220}Rn) in addition to measuring soil permeability, temperature, humidity, and other technical parameters that indicate the proper operation of the instrument. The system utilizes silicon radiation detectors in the measurement chamber for alpha spectroscopy, which separates the different radon daughter products (Baskaran 2016).

The concentration of radon (^{222}Rn) gas is determined by measuring the short-lived daughter products produced by radon decay. The number of ^{218}Po ions collected on a semiconductor detector is directly proportional to the radon gas concentration in the instrument's chamber. The half-life of ^{218}Po is 3.05 min, and the equilibrium between the radon decay rate and ^{218}Po detector activity is achieved after approximately 15 min, which is the minimum achievable response time (Cvetković et al. 2021).

2.6 Radon Measurement in Water

When collecting water samples for radon analysis, it is important to take extra care as radon is a diffusive gas. The standard procedures for collecting dissolved gases in water can be applied to radon as well. The emanation method is one of the oldest techniques used for determining radon activity in water (Lucas 1957). This involves circulating air or a carrier gas in a closed loop to extract dissolved radon from the water, and the emanated radon gas is measured using a suitable radon detector (Lee and Burnett 2013; Peterson et al. 2013).

The emanation method is commonly used for alpha counting of radon and its daughter products. The alpha, beta, and gamma activities of radon daughters can also be directly measured by counting the water sample. Since the radon activity in groundwater is generally two to three orders of magnitude greater than that of surface water, a small quantity of water sample is sufficient for measuring radon in groundwater (Sukanya et al. 2022).

2.6.1 Measurement of Radon in Groundwater

Groundwater samples ranging from 10 to 250 mL can be collected in glass bottles for radon measurement, depending on the type of radon detector used. Careful attention should be paid to sample collection and analysis to ensure accurate results. The collected samples can be measured in the laboratory within 3–4 days of sample collection using a closed-circuit method. In this method, air is bubbled through the water sample, and the radon activity in the gas phase is determined as shown in Fig. 2.1.

It is recommended that at least four times the air volume should be passed through the bubbler to attain equilibrium of radon between water and air, which typically occurs within 5 min (Óskarsson and Ásgeirsdóttir 2017). To achieve acceptable precision, the radon activity should be counted for 40 min (4 cycles of counting) after reaching equilibrium.

$$C_{\text{water}} = C_{\text{gas}}\left(\frac{V_{\text{gas}}}{V_{\text{water}}} + \alpha(T)\right) \tag{2.1}$$

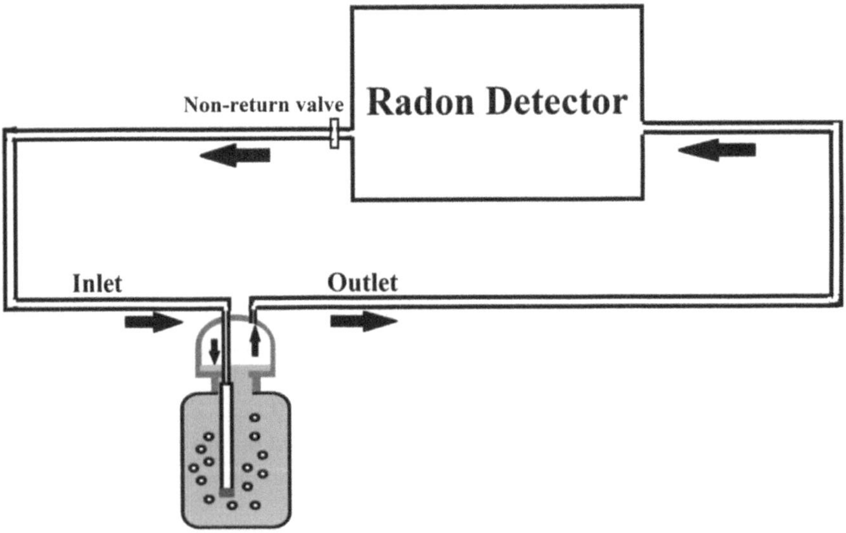

Fig. 2.1 Measurement of dissolved radon in groundwater by bubbling air in a closed loop. (*Source* Sukanya et al. 2022)

where V_{gas} and V_{water} denote the volumes of the air and water in the closed loop, respectively, and $\alpha(T)$ is the temperature-dependent partition coefficient, also known as "Ostwald solubility" of radon. An equation for $\alpha(T)$ is provided by Weigel (1978):

$$\alpha(T) = 0.105 + 0.405\,\mathrm{e}^{-0.0502\times T}\ (T \text{ in } ^\circ\mathrm{C}) \tag{2.2}$$

At room temperature, $\alpha(T)$ is about 0.25, indicating that the volumetric radon activity in the gas phase is four times higher than in the water phase.

Radon decay correction is crucial in accurately measuring its activity using detectors. The scintillation cell method stands out as a cost-effective and highly sensitive means of measuring radon activity in water. Among the various methods for measuring radon activity in water, the scintillation cell method stands out as highly sensitive and cost-effective. With a sample volume requirement of only 20 mL and a buildup time of 3 h for all alpha-emitting radon daughter products, this method can provide accurate measurements.

Moreover, the well-known instrument response function enables immediate measurement after bubbling. The buildup and decay of the total alpha activity can be approximated by a single exponential with a half-life of 40 min (Surbeck 2007). The scintillation cell method is highly specific and sensitive, functioning even at 100% relative humidity. With a counting time of 60 min, this method can detect low radon activities, with detection limits of 1.0 Bq/m^3 for a sample volume of 5 L and 25 Bq/m^3 for a sample volume of 20 mL (Stringer and Burnett, 2004).

For the purpose of radon detection in water samples, electronic detectors that are either semiconductor detectors or ionization chambers are also employed. To obtain adequate sensitivities, those using semiconductor detectors need a large sample volume (250 mL) and an accumulation of the daughter products on the detector by a high voltage. According to L'Annunziata (2012), one benefit of this technology is that the detector not only counts but also has a good enough energy resolution to distinguish between various radon progenies.

The requirement to dry radon gas before it enters the sensor is a drawback for the majority of semiconductor detectors. Anhydrite (dehydrated calcium sulfate), is commonly used as a desiccant, however, silica gel is not suitable for this purpose due to its radon adsorption property. Typically, outgassing 40 ml samples allow for a detection limit of about 1000 Bq/m^3, 250 ml sample for around 400 Bq/m^3 for 10 min counting in semiconductor detectors (RAD7 User Manual Durridge 2016).

Using LSC, direct groundwater sample counting only needs a very small sample size (10 mL). Although this approach lacks specificity, it is very sensitive. i.e., any alpha or beta emitters that are soluble in the scintillator except radon and its progeny also contribute to the count rate. The radon detection threshold using LSC is 370 Bq/m^3 (Crawford-Brown 1990). The benefit of direct gamma counting is the lack of sample pretreatment. However in order to test radon, a comparatively larger volume of water sample (1.3 L) is required. For a sample of 1 L of water, the detection limit for the gamma counting method is 370 Bq/m^3 (Yang 2020).

2.6.2 *Measurement of Radon in Surface Water*

The in-situ monitoring of radon in surface water is critical due to its low activity and short half-life. This requires the continuous pumping of surface water using a peristaltic pump and its jetting into an air–water exchanger for radon stripping (Fig. 2.2). The stripped radon is circulated via a closed air loop, including a desiccant tube, into a ^{222}Rn counting system (Peterson et al. 2013; De Simone et al. 2015). Equilibration between the liquid and gas phases occurs within 30 min. To ensure acceptable precision, the radon activity at each location should be counted for 2 hours (3 cycles of counting) after reaching equilibrium.

The measured gas phase radon activity concentration needs to be converted to radon activity in water using Eq. (2.1). The semiconductor detector-based method employed for this approach allows for a lower limit of radon measurement of ± 1 Bq/m^3 with a precision of $\pm 10\%$. While this method provides in situ monitoring, it requires the use of a peristaltic pump, and the accuracy can be affected by temperature, pressure, and humidity changes during sampling.

2.7 Detector Suitability for Radon Measurement

The choice of detectors and sample techniques for radon measurement relies on the specific goal of the investigation. For instance, scintillation counters employ active method with grab or continuous sampling (Bem et al. 2020). Both active

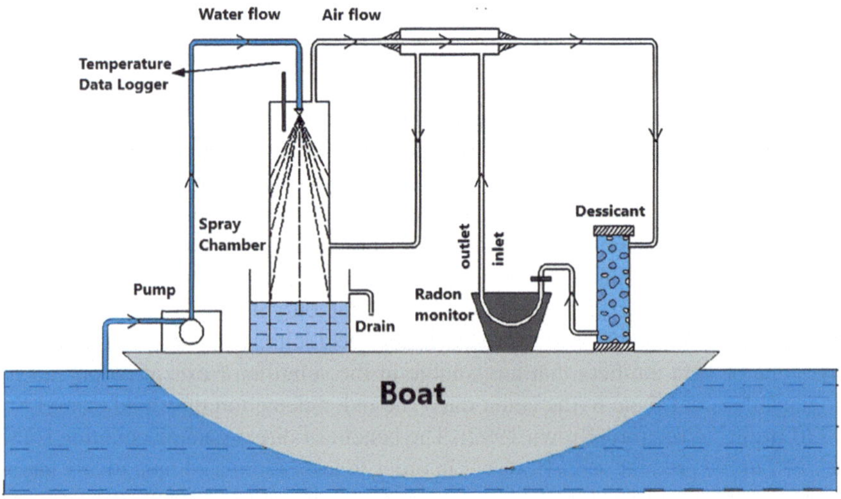

Fig. 2.2 Schematic sketch of in situ ^{222}Rn monitoring system for surface water measurements (*adapted from* Sukanya et al. 2022)

(one-time cycle grab sampling) and passive (repeated cycle continuous sampling) approaches can be used in an ionization chamber. In contrast to passive method, which is appropriate for diffusion chambers with semi conductor detectors, active method is applied in semiconductor detectors. SSNTD can be treated with a passive technique using one-time cycle continuous sampling (Janik et al. 2014). SSNTD, however, is ineffective for determining the level of radon activity in water.

2.8 Comparing Radon-In-Water Measurement Techniques

This section provides a comprehensive comparison and summary of various techniques for measuring radon in water. Table 2.1 presents a comparison of dissolved radon (in water) measurement detectors based on the literature, with additional information provided in Table 2.2.

The presented sensitivity range and uncertainty are generalizations and can slightly vary for different commercially available radon detectors, depending on the device manufacturer and model. The Lucas cell method is still the most sensitive

Table 2.1 Comparative overview of radon detector specifications (*Source* Sukanya et al. 2022)

Detector type	Lucas cell	Silicon detector	[a]IC	[a]LSC	[a]GS
Sensitivity range	0.8–16 cph/Bq/m^3	0.02 cph/Bq/m^3	1.05–3 Bq/m^3 h	No data	No data
Sampling volume	0.02–3 L	~0.27 L	0.5 L	0.01–2 L	0.1–2 L
Uncertainty	<10%	5–12%	5–12%	~10%	5–13%
Counting efficiency	15–90%	95%	No data	280–300%	1.5%
Lowest detection limit	0.3–1 Bq/L	0.04–0.4%	0.3 Bq/L	0.05–1 Bq/L	10 Bq/L

[a]*IC* Ionization Chamber, *LSC* Liquid Scintillation Counter, *GS* Gamma-ray Spectrometer

Table 2.2 Summary of sample requirements and turnaround time of various techniques for radon in water (*Source* Sukanya et al. 2022)

Detector	Sample volume (L)	Sample treatment	Turnaround time (h)
Gamma-ray spectrometry	0/1–2	No	4–13
Ionization chamber	0.05–0.2	Degassing	<1
LSC	0.01–2	Dependent on cocktail	3–8
Scintillation/Lucas cells	0.05–1	Degassing	1–3
Silicon detector	0.1–0.4	Degassing	<1

technique among various electronic detectors, with a low sample volume requirement (20 mL) and a low price. The method has a well-known instrument response function and takes 3 h for all alpha-emitting radon daughter products to build up.

The advantage of the Lucas cell method is its specificity, high sensitivity, and ability to work well even at 100% relative humidity. This method can measure samples with very low radon activities (detection limit—1.0 Bq/m^3 for a sample volume of 5 L and 25 Bq/m^3 for a sample volume of 20 mL with a counting time of 60 min) (Stringer and Burnett 2004).

Semiconductor detectors or ionization chambers-based electronic detectors are also commonly used for radon detection in water samples. The semiconductor detectors require a large sample volume (250 mL) and an accumulation of daughter products on the detector by applying a high voltage to obtain reasonable sensitivities. This method provides not only counting but also good energy resolution to distinguish between different radon progenies (L'Annunziata 2012).

A disadvantage of this method is that the radon gas needs to be dried before it enters the instrument, which can be done with a drierite cartridge. Commercially available semiconductor detectors (e.g., RAD7 radon detector) can provide faster results and are suitable for onsite field measurements. Ionization chamber technique offers high precision, ruggedness, ease of operation, and other advantages. However, it is sensitive to background gamma radiations, which require background corrections (Tracy 2010).

In Liquid Scintillation Counting (LSC) technique, only small amount of groundwater sample (10 mL) is required. However, this method lacks specificity as any alpha or beta emitters (other than radon and its daughters) that are soluble in the scintillator also contribute to the count rate. Direct gamma counting does not require sample preparation, which is an advantage. However, a relatively larger volume of water sample (1.3 L) is necessary for radon measurement (Sukanya et al. 2022).

2.9 RAD7 Instrument—Radon in Water

The RAD7 radon detector with a RAD-H_2O accessory (Durridge, U.S.A) is widely used for efficient radon detection in water. This equipment offers a fast and inexpensive on-site analysis of radon in water, making it an ideal tool for field measurements. The detector is compact, portable, and powered by a battery, while the RAD-H_2O provides results in just 30 min. The semiconductor detector method employed by this instrument involves a silicon semiconductor detector mounted in a cell that detects alpha particles emitted from radon and thoron progenies that are electrostatically collected on the detector surface. This simple and safe method eliminates the need for harmful chemicals during the analysis process (RAD7 User Manual Durridge 2016).

The detector is equipped with a bubbling kit that facilitates the closed-loop degassing of radon from a water sample into the air, which is then measured to determine the radon concentration in the water. The setup includes a 250 mL radon-tight reagent container with a zinc sulfide-coated detection chamber that serves as a scintillator for detecting alpha activity. Additionally, a glass bulb containing calcium chloride is used as an absorbent to remove moisture. After pumping the air in a closed circuit for 5–10 min to mix the radon and air, the resultant alpha activity is recorded to provide an immediate measurement of the radon concentration in the water sample.

To ensure accurate results, it is crucial to handle the water sample with great care and avoid any contact with air. The recommended procedure involves allowing the water to flow through a 250 mL vial for a period of time, ensuring that the vial remains filled and continuously flushed with water. Once filled, the vial should be capped while still submerged in the water. It is important to eliminate any bubbles in the vial and ensure that the cap is tightened securely.

2.9.1 RAD-H_2O Modus Operandi

This technique includes air purging to remove residual radon from the system and keep the RAD7 detector dry. This is achieved by flushing the system with clean air using the built-in pump for at least 10 min before each measurement.

To further reduce moisture inside the system, a large drying unit is used in a closed loop with the RAD7 detector. This process can take up to 30 min until the relative humidity inside the RAD7 is less than 10%. This step minimizes errors caused by moisture during the measurement process.

During sample reading, small drierite tubes are attached to the system. For water sample analysis, an outlet tube from the RAD7 is connected to an aerator cap to provide aeration. A glass frit is placed below the aerator cap to produce bubbles, allowing radon to evenly degas from the water sample. The radon activity concentration value appears on the digital display integrated into the system after the bubbling process.

2.9.2 The Closed-Loop Concept

The RAD H_2O method employs a closed-loop aeration system that maintains constant volumes of water and air, independent of the flow rate. As air circulates through the water, radon is extracted until an equilibrium point is reached, which takes about 5 min. At this point, no more radon can be extracted from the water, and the extraction efficiency is typically 99% for a 40 mL sample and 94% for a 250 mL sample. If samples are not analyzed immediately, the radon concentration in the sample will decline due to radioactive decay. Thus, decay correction must be applied, which is a simple exponential function with a time constant of 132.4 h. The decay correction factor (DCF) is given by exp (T/132.4), where T is the decay time in hours.

2.10 Quality Assurance of Measurement

To ensure accurate and precise data in radon measurement, quality assurance (QA) is critical (IAEA 2020). Four types of quality control measurements should be followed to achieve QA, including calibration measurements, background measurements, duplicate/collocated measurements, and routine instrument performance checks. These measurements have been detailed by various authors (Hager et al. 2017; Pantelić et al. 2019; Tamakuma et al. 2021; Radulescu et al. 2022).

2.10.1 Calibration Measurements

Calibration measurements should be performed periodically for continuous monitors by the manufacturer or an authorized reference laboratory trained by the manufacturer (ISO 2019; Mas et al. 2021). During calibration, several steps should be taken, including checking and adjusting voltage/wave patterns at critical points in the circuitry, checking and recharging batteries, determining proper discriminator settings, determining background by exposure to a radon-free environment of nitrogen or aged air, and checking calibration factor by exposure to a reference atmosphere (Radulescu et al. 2022).

2.10.2 Background Measurements

Conducting an annual assessment of background is crucial for continuous monitors and passive detectors to ensure the accuracy of the measurement (Röttger et al. 2021). The build-up of alpha-particle emitting ^{210}Pb over time and the residual background from neutron emissions can increase the background level, which can limit the sensitivity of the detector (Pelczar et al. 2021). If the measured background is below 30 Bq/m^3, the interval between each background measurement can be increased, except when the system is exposed to high activities of radon in a sample (Zeeb et al. 2009).

2.10.3 Duplicate/Collocated Measurements

To ensure the accuracy of radon measurements, it is recommended to duplicate measurements in about 10% of the total number of measurement locations. This can be achieved by colocating two radon monitors side by side, as suggested by experts in the field (Zeeb et al. 2009; Barros et al. 2015). The precision of the detector can then be evaluated by calculating the relative percent difference between the two measurements. If the measurements fall outside the expected precision range, further investigation is necessary to identify the issue.

To maintain the reliability of radon monitors, duplicate measurements should be conducted at regular intervals, such as every 6 months, or periodically between calibrations, as recommended by Zeeb et al. (2009) and Hofmann et al. (2012). These measures not only ensure the quality of data but also help to identify any potential sources of error or bias in the measurement process, which is critical for accurate radon risk assessment and management.

2.10.4 Routine Instrument Performance Checks

Regular performance checks are critical to ensure accurate and reliable results from radon monitors. Such checks should be carried out systematically on a daily basis by the user. Key components of the instrument, including the pump, battery (to ensure the minimum voltage required for proper detector functioning), and electronics, should be routinely checked and the results should be logged. In the case of radon monitors that use semiconductor detectors, regular humidity correction is necessary to maintain optimal performance (Galli et al. 2019). By implementing these measures, users can ensure that their radon monitoring instruments remain in optimal state and produce reliable data over the long term.

2.10.5 Proficiency Tests and Inter-Laboratory Comparisons

Participation in proficiency tests and inter-laboratory comparisons is crucial for quality control, as it helps to standardize and improve technical measurement methods (Janik et al. 2014; Berlier et al. 2019). These activities help to standardize technical measurement methods and improve overall quality assurance. The Radiation Protection Division of Health Protection Agency (HPA, UK) and the National Institute of Radiological Science (NIRS, Japan) are two organizations that conduct well-known worldwide periodic exercises, ideally on an annual or biennial basis, and publish the results.

These exercises are typically conducted in a walk-in radon chamber, where radon activity concentrations, along with temperature, humidity, air pressure, etc., can be measured and compared. Performance quality of laboratories has been found to be improving based on international intercomparison data of radon measurements carried out at NIRS. Parameters such as Percent Difference (PD), normalized error (En), standard deviation, z-score, Percentage Precision Error (PER), and others are commonly used for comparisons. However, it is important to note that in-field intercomparison exercises should also be conducted regularly, as real-life scenarios where radon monitors (active or passive) are installed may be complex and differ from the controlled stable conditions used for testing (Berlier et al. 2019; Janik et al. 2014; Rábago et al. 2020, 2022).

2.11 Conclusion and Future Research

To ensure reliable and precise environmental data, selecting appropriate measurement techniques and understanding analytical methodology are crucial. However, most of the literature on radon sampling and analysis is scattered and diverse, making it challenging to choose appropriate measurement techniques. In this chapter, we offer a comprehensive overview of the fundamentals of radon sampling and analysis, with a focus on measurement approaches for radon in water, which are particularly relevant for hydrological applications. By delving into the analytical methodology and selection of appropriate techniques, we aim to provide readers with the knowledge needed to ensure reliable and precise measurements of radon in various environmental contexts.

Radon measurement in water demands a meticulous approach with the selection of appropriate techniques being of paramount importance. A plethora of methods are available, such as the Lucas cell, ionization chamber, semiconductor detectors, gamma-ray spectrometry, and LSC, to name a few. A comparison of these methods demonstrates that the Lucas cell approach, due to its high precision and accuracy, is the most sensitive and reliable. However, for on-site field measurements with quick results, commercially available semiconductor detectors perform effectively. In addition to methodology, quality assurance is a vital aspect to ensure that the obtained radon data is reliable, precise, and accurate. With this comprehensive overview, we aim to provide an in-depth understanding of the fundamentals of radon sampling and analysis, with a particular emphasis on measurement approaches for radon in water, which is critical for hydrological applications.

To gain a better picture regarding the dynamics of hydrogeological processes under varying temporal conditions, it is crucial to develop hybrid/altered instruments that can provide higher precision and accuracy. One promising approach involves the use of energy-efficient instruments with cloud-based servers for continuous measurement and real-time monitoring, enabling a significant improvement in the quality of data obtained. However, despite the availability of many reliable techniques, future research should prioritize the development of AI-based smart detectors that can identify and rectify uncertainties and errors in real time, thereby advancing our knowledge of hydrogeological systems.

References

Abdalla AM, Al-Hajry A (2015) Radon irradiation chamber and its applications. Nucl Instrum Methods Phys Res, Sect A 786:78–82

Abdalla AM, Ali AM, Al-Jarallah M, Okada G, Kawaguchi N, Yanagida T (2019) Radon detection using alpha scintillation KACST cell. Nucl Instrum Methods Phys Res, Sect A 922:84–90

Abdalla AM, Al-Naggar TI, Bashiri AM (2021) Radon gas build up using alpha scintillation cell: experimentally and theoretically. Appl Radiat Isot 175:109796

Abdalla AM, Ismail AM, Al-Naggar TI (2017) Radon calibration system. In: Radon. IntechOpen

Abdalla AM, Siddiqui SA, Kawaguchi N, Yanagida T (2022) An exhaustive study of the efficiency and sensitivity of a radon cell coated with a new scintillation material. Nucl Instrum Methods Phys Res, Sect A 1025:166165

Álvarez V, Borges FIG, Cárcel S, Cebrián S, Cervera A, Conde CAN, Dafni T, Díaz J, Egorov M, Esteve R, Evtoukhovitch P (2013) Ionization and scintillation response of high-pressure xenon gas to alpha particles. J Instrum 8(05):P05025

Antonopoulos-Domis M, Xanthos S, Clouvas A, Alifrangis D (2009) Experimental and theoretical study of radon distribution in soil. Health Phys 97(4):322–331

Banjanac R, Dragi´c A, Grabeˇz B, Jokovi´c D, Markushev D, Pani´c B, Udovicic V (2006) Indoor Radon measurements by nuclear track detectors: applications in secondary schools. Facta Univ Ser Phys Chem Technol 4:93–100

Banjanac R, Dragić A, Grabež B, Joković D, Markushev D, Panić B, Udovičić V, Aničin I (2006) Indoor radon measurements by nuclear track detectors: applications in secondary schools. Facta Univ Ser Phys Chem Techn 4(1):93–100

Barros N, Field DW, Steck DJ, Field RW (2015) Comparative survey of outdoor, residential and workplace radon concentrations. Radiat Prot Dosimetry 163(3):325–332

Baskaran M (2016) Radon: A tracer for geological, geophysical and geochemical studies (Vol. 367) Springer, Basel

Bem H, Długosz-Lisiecka M, Janiak S, Mazurek D, Szajerski P (2017) Fast determination of indoor radon (222 Rn) concentration using liquid scintillation counting. J Radioanal Nucl Chem 312:337–342

Bem H, Długosz-Lisiecka M, Mazurek-Rudnicka D, Szajerski P (2021) Occurrence of 222Rn and 226,228 Ra in underground water and 222Rn in soil and their mutual correlations for underground water supplies in southern Greater Poland. Environ Geochem Health 43(8):3099–3114

Bem H, Gasiorowski A, Szajerski P (2020) A fast method for the simultaneous determination of soil radon (222Rn) and thoron (220Rn) concentrations by liquid scintillation counting. Sci Total Environ 709:136127

Berlier F, Cardellini F, Chiaberto E, Garlati L, GIUFFRIDA D, Ragani MF, Leonardi F, Magnoni M, Minchillo G, Prandstatter A, Serena E (2019) Main results of the second AIRP international radon-in-field intercomparison for passive measurement devices. Radiat Meas 128:1–11

Blanco-Novoa O, Fernández-Caramés TM, Fraga-Lamas P, Castedo L (2018) A cost-effective IoT system for monitoring indoor radon gas concentration. Sensors 18(7):2198

Bonczyk M, Samolej K (2019) Testing of the radon tightness of beakers and different types of sealing used in gamma-ray spectrometry for ^{226}Ra concentration determination in NORM. J Environ Radioact 205:55–60

Bučar K, Korun M, Vodenik B (2012) Influence of the thorium decay series on the background of high-resolution gamma-ray spectrometers. Appl Radiat Isot 70(6):1005–1009

Cahyadi L, Pangaribuan J, Saputra KVI (2019) Determination of environmental condition in universitas Pelita Harapan based on radon concentration analysis. J Phys Conf Ser 1245(1):012073. IOP Publishing

Celaya S, Fuente I, Quindós L, Sainz C (2018) Optimization of a portable liquid scintillation counting device for determining 222Rn in water. Radiat Meas 117:1–6

Crawford-Brown DJ (1990) Analysis of the health risk from ingested radon. Radon, radium and uranium in drinking water. Lewis Publishers, Chelsea, Michigan, pp 17–26

Cvetković M, Kapuralić J, Pejić M, Kolenković Močilac I, Rukavina D, Smirčić D, Kamenski A, Matoš B, Špelić M (2021) Soil gas measurements of radon, CO_2 and hydrocarbon concentrations as indicators of subsurface hydrocarbon accumulation and hydrocarbon seepage. Sustainability 13(7):3840

De Simone G, Galli G, Lucchetti C, Tuccimei P (2015) Calibration of big bottle RAD H_2O set-up for radon in water using HDPE bottles. Radiat Meas 76:1–7

Dearnaley G, Whitehead AB (1961) The semiconductor surface barrier for nuclear particle detection. Nucl Inst Methods 12:205–226

Di Carlo C, Ampollini M, Antignani S, Caprio M, Carpentieri C, Bochicchio F (2022) Thoron interference on performance of continuous radon monitors: an experimental study on four devices and a proposal of an indirect method to estimate thoron concentration. Int J Environ Res Public Health 19(4):2423

Dimitrova I, Georgiev S, Mitev K, Todorov V, Dutsov C, Sabot B (2023) Study of the performance and time response of the RadonEye Plus2 continuous radon monitor. Measurement 207:112409

do Nascimento Santos JM, dos Santos Júnior JA, dos Santos Amaral R, Bezerra MBCF, e Fernandes TDS, Rojas LAV, Fernández ZH, de Araújo EEN, de Araújo Rodrigues AJI, de Barros Correia FL, Milán MO (2021) α Espec-222 as a variant to the radon monitoring method using a solid state nuclear track. Appl Radiat Isot 178:109948

Durrani SA, Bull RK (2013) Solid state nuclear track detection: principles, methods and applications, vol 111. Elsevier

El Ghazaly M, Aydarous A, Salama TT, El-Naggar HI (2021) Characterization of Makrofol® DE 1-1 (bisphenol-A polycarbonate) as SSNTD and its application in the alpha particle detection. Radiat Phys Chem 180:109101

Engelbrecht R (2020) Environmental radioactivity monitoring. In Handbook of Radioactivity Analysis 2:1–40 Academic Press, Netherlands

EPA (US Environmental Protection Agency) (1992) Indoor radon and radon decay product measurement device protocols. Office of Air and Radiation (6604J), EPA 402-R-92-004 July (revised)

Frutos-Puerto S, Hurtado-Sanchez MC, de la Torre Pérez J, Pinilla-Gil E, Miró C (2021) Radon alpha track counting on solid state nuclear track detector by an ImageJ-based software macro. Appl Radiat Isot 173:109695

Galli G, Cannelli V, Nardi A, Piersanti A (2019) Implementing soil radon detectors for long term continuous monitoring. Appl Radiat Isot 153:108813

Geiger HJ (1967) The neighborhood health center: education of the faculty in preventive medicine. Arch Environ Health: Int J 14(6):912–916

Gopalakrishnan P, Jeyanthi J (2022) Importance of radon assessment in indoor environment—a review. Mater Today: Proceed 56(3):1495–1500

Gubanski A, Kupracz J, Kostyla P, Kaczorowska D, Wrobel J (2019) Application of the electret in alpha radiation sensor to measure the concentration of radon in selected ambient conditions. J Sens

Hager L, Tanner R, Gilvin P, Eakins J, Baker S (2017) The impacts of a new electrochemical etch cycle for the Public Health England neutron personal dosimetry service. Radiat Meas 106:303–311

Higuchi S, Kamishiro Y, Ishihara M, Yasuoka Y, Mori Y, Hosoda M, Iwaoka K, Tokonami S, Takahashi R, Janik M, Muto J (2019) Evaluation of a radon air monitor in the measurement of radon concentration in water in comparison with a liquid scintillation counter. Radiat Prot Dosim 184(3–4):426–429

Hofmann W, Arvela HS, Harley NH, Marsh JW, McLaughlin J, Röttger A, Tokonami S (2012) Appendix A radon and radon progeny metrology and quality assurance of measurements. J ICRU 12(2):157–160

Horrocks DL, Studier MH (1964) Determination of radioactive Nobel gases with a liquid scintillator. Anal Chem 36(11):2077–2079

Hou X (2018) Liquid scintillation counting for determination of radionuclides in environmental and nuclear application. J Radioanal Nucl Chem 318(3):1597–1628

IAEA (2020) Design and conduct of indoor radon surveys. Saf Rep Ser No 98

Irlinger J, Trinkl S, Wielunksi M, Tschiersch J, Rühm W (2016) Monte Carlo simulation of semiconductor detector response to 222Rn and 220Rn environments. J Environ Radioact 158:64–70

ISO 11665-1 (2019) Measurement of radioactivity in the environment—air: radon-222—Part 1: origins of radon and its short-lived decay products and associated measurement methods

ISO (2019) Measurement of radioactivity in the environment—air: radon-222—Part 1: origins of radon and its short-lived decay products and associated measurement methods. ISO 11665-1

Janik M, Ishikawa T, Omori Y, Kavasi N (2014) Invited article: radon and thoron intercomparison experiments for integrated monitors at NIRS, Japan. Rev Sci Instrum 85(2):022001

Jobbágy V, Altzitzoglou T, Malo P, Tanner V, Hult M (2017) A brief overview on radon measurements in drinking water. J Environ Radioact 173:18–24

Jobbágy V, Stroh H, Marissens G, Hult M (2019) Comprehensive study on the technical aspects of sampling, transporting and measuring radon-in-water. J Environ Radioact 197:30–38

Joshirao PM, Shin JW, Vyas CK, Kulkarni AD, Kim H, Kim T, Hong SW, Manchanda VK (2013) Development of optical monitor of alpha radiations based on CR-39. Appl Radiat Isot 81:184–189

Korun M, Petrovič T, Vodenik B, Zorko B (2023) Influence of the solar activity on the background of a high-resolution gamma-ray spectrometer. Appl Radiat Isot 110683

Korun M, Vodenik B, Zorko B (2014) Determination of the shielding factors for gamma-ray spectrometers. Appl Radiat Isot 87:372–375

Korun M, Vodenik B, Zorko B (2016) Measurement function for the activities of multi-gamma-ray emitters in gamma-ray spectrometric measurements. Appl Radiat Isot 109:518–521

L'Annunziata MF (ed) (2012) Handbook of radioactivity analysis. Academic Press

L'Annunziata MF (ed) (2020) Handbook of radioactivity analysis: volume 2: radioanalytical applications. Academic Press

Laubenstein M, Lawson I (2020) Low background radiation detection techniques and mitigation of radioactive backgrounds. Front Phys 8:577734

Lee KY, Burnett WC (2013) Determination of air-loop volume and radon partition coefficient for measuring radon in water sample. J Radioanal Nucl Chem 298:1359–1365

Lucas HF (1957) Improved low-level alpha-scintillation counter for radon. Rev Sci Instrum 28(9):680–683

Maier A, Jones J, Sternkopf S, Friedrich E, Fournier C, Kraft G (2021) Radon adsorption in charcoal. Int J Environ Res Public Health 18(9):4454

Malinovsky G, Yarmoshenko I, Vasilyev A (2019) Meta-analysis of case–control studies on the relationship between lung cancer and indoor radon exposure. Radiat Environ Biophys 58(1):39–47

Mas JL, Velázquez LM, Hurtado-Bermúdez SJ (2021) Optimization of AlphaGUARD AquaKIT set-up for analysis of radon in water using stainless-steel bottles and its validation through IAEA standard samples. Radiat Meas 142:106545

McColl N, Auvinen A, Kesminiene A, Espina C, Erdmann F, de Vries E, Greinert R, Harrison J, Schüz J (2015) European Code against cancer 4th edition: Ionising and non-ionising radiation and cancer. Cancer Epidemiol 39:S93–S100

Mertes F, Röttger S, Röttger A (2020) A new primary emanation standard for Radon-222. Appl Radiat Isot 156:108928

Morales-Simfors N, Wyss RA, Bundschuh J (2020) Recent progress in radon-based monitoring as seismic and volcanic precursor: a critical review. Crit Rev Environ Sci Technol 50(10):979–1012

National Council on Radiation Protection and Measurements (NCRP) (1988) Measurements in radon and radon daughters in air. NCRP Report 97. NCRP Publications, Bethesda, MD. http://www.ncrppublications.org/Reports/097

Óskarsson F, Ásgeirsdóttir RS (2017) Radon in Icelandic cold groundwater and low-temperature geothermal water. Procedia Earth Planet Sci 17:229–232

Pantelić G, Čeliković I, Živanović M, Vukanac I, Nikolić JK, Cinelli G, Gruber V (2019) Qualitative overview of indoor radon surveys in Europe. J Environ Radioact 204:163–174

Papastefanou C (2002) An overview of instrumentantion for measuring radon in soil gas and groundwaters. J Environ Radioact 63(3):271–283

Papastefanou C (2007) Measuring radon in soil gas and groundwaters: a review. Ann Geophys

Pelczar K, Zuzel G, Wójcik M, Pocar A, Ianni A (2021) An online radon monitor for low-background detector assembly facilities. The Eur Phys J C 81(1):1–12

Pérez B, López ME, Palacios D (2021) Concentric rings formation on etched LR-115 in bare mode by unconventional exposition. Nucl Instrum Methods Phys Res, Sect B 496:37–44

Peterson RN, Breier JA, Harmon LR, Brusa J, Hutchins PR (2013) Development of a sparging chamber for field radon analysis. J Radioanal Nucl Chem 298:1347–1357

Pujol L (2022) Validation of a method for measuring radon-222 in water using two-phase liquid scintillation counting according to the ISO/IEC 17025 criteria. J Radioanal Nucl Chem 1–10

Rábago D, Fuente I, Celaya S, Fernandez A, Fernandez E, Quindos J, Pol R, Cinelli G, Quindos L, Sainz C (2020) Intercomparison of indoor radon measurements under field conditions in the framework of metroradon european project. Int J Environ Res Public Health 17(5):1780

Rábago D, Quindós L, Vargas A, Sainz C, Radulescu I, Ioan MR, Cardellini F, Capogni M, Rizzo A, Celaya S, Fuente I (2022) Intercomparison of radon flux monitors at low and at high radium content areas under field conditions. Int J Environ Res Public Health 19(7):4213

RAD7 User Manual Durridge (2016) Durridge radon instrument. Durridge Co, USA

Radulescu I, Calin MR, Luca A, Röttger A, Grossi C, Done L, Ioan MR (2022) Inter-comparison of commercial continuous radon monitors responses. Nucl Instrum Methods Phys Res, Sect A 1021:165927

Röttger A, Röttger S, Grossi C, Vargas A, Curcoll R, Otáhal P, Hernández-Ceballos MÁ, Cinelli G, Chambers S, Barbosa SA, Ioan MR (2021) New metrology for radon at the environmental level. Meas Sci Technol 32(12):124008

Ryu J, Lee HW, Chae JS (2022) Rapid sequential determination of 222Rn and 226Ra in drinking water by liquid scintillation counting. Talanta 239:123141

Saha U, Turner P, Cooper D, Kitto M (2021) The AARST consortium standard on protocol for the collection, transfer and measurement of radon in water—an update (No. IAEA-CN--279)

Salonen L, Kaihola L, Carter B, Cook GT, Passo CJ Jr (2012) Environmental liquid scintillation analysis. In: Annunziata MF (ed) Handbook of radioactivity analysis. Academic Press, pp 626–678

Sethy NK, Jha VN, Ravi PM, Tripathi RM (2014) A simple method for calibration of Lucas scintillation cell counting system for measurement of ^{226}Ra and 222Rn. J Radiat Res Appl Sci 7(4):475–477

Seo J, Kim G (2021) Rapid and precise measurements of radon in water using a pulsed ionization chamber. Limnol Oceanogr: Methods 19(4):245–252

Sorimachi A, Nagamatsu Y, Omori Y, Ishikawa T (2021) Comparison experiments for radon and thoron measuring instruments at low-level concentrations in one room of a Japanese concrete building. Appl Radiat Isot 173:109696

Steinhauser G, Buchtela K (2020) Gas ionization detectors. In: Handbook of radioactivity analysis. Academic Press, pp 245–305

Stojković I, Todorović N, Nikolov J, Tenjović B (2016) PSA discriminator influence on 222Rn efficiency detection in waters by liquid scintillation counting. Appl Radiat Isot 112:80–88

Stringer CE, Burnett WC (2004) Sample bottle design improvements for radon emanation analysis of natural waters. Health Phys 87(6):642–646

Studnička F, Štěpán J, Šlégr J (2019) Low-cost radon detector with low-voltage air-ionization chamber. Sensors 19(17):3721

Sukanya S, Noble J, Joseph S (2022) Application of radon (222Rn) as an environmental tracer in hydrogeological and geological investigations: an overview. Chemosphere 135141

Surbeck H (2007) Dissolved gases as natural tracers in karst hydrogeology; radon and beyond. Center of Hydrogeology (CHYN), University of Neuchâtel, Emile-Argand, p 11

Syuryavin AC, Park S, Nirwono MM, Lee SH (2020) Indoor radon and thoron from building materials: analysis of humidity, air exchange rate, and dose assessment. Nucl Eng Technol 52(10):2370–2378

Tamakuma Y, Kranrod C, Jin Y, Kobayashi H, Nugraha ED, Sanpei A, Kiso M, Hashimoto H, Negami R, Hosoda M, Tokonami S (2021) Characterization of commercially available active-type radon-thoron monitors at different sampling flow rates. Atmosphere 12(8):971

Tchorz-Trzeciakiewicz D, Kozłowska B, Walencik-Łata A, Podstawczyńska A, Olszewski J, Bartak J, Karpińska M, Wołoszczuk K, Dohojda M, Nowak J, Długosz-Lisiecka M (2020) Radon intercomparison tests Stanisław Chałupnik, Krystian Skubacz, Małgorzata Wysocka, Jadwiga Mazur,–Katowice, 2016 Michał Bonczyk, Krzysztof Kozak, Dominik Grządziel. Paweł Urban. Nukleonika 65(2):127–132

Tokonami S, Pan J, Matsumoto M, Furukawa M, Fujimoto K, Fujitaka K, Kurosawa R (1996) Radon measurements in indoor workplaces. Radiat Prot Dosimetry 67(2):143–146

Tracy BL (2010) In: Radon Atwood DA (Ed) Radionuclides in the environment. Wiley, pp 191–206

US EPA (1992) National radon survey. Summary Report EPA 402-R-92-01

Vajda N, Pöllänen R, Martin P, Kim CK (2020) Alpha spectrometry. In: Handbook of radioactivity analysis. Academic Press, pp 493–573

Van Dung N, Thuan DD, Nhan DD, Carvalho FP, Van Thang D, Quang NH (2022) Radiation exposure in a region with natural high background radiation originated from rare earth element deposits at Bat Xat district, Vietnam. Radiat Environ Biophys 61(2):309–324

Venkataraman R (2020) Semiconductor detectors. In: Handbook of radioactivity analysis. Academic Press, pp 409–491

Warkentin P, Curry E, Michael O, Bjorndal B (2020) A comparison of consumer-grade electronic radon monitors. J Radiol Prot 40(4):1258

Weigel F (1978) Radon. Chem Ztg 102(9):287–299

Yang IC (2020) Sampling and analysis of dissolved radon-222 in surface and ground water. In: Radon, radium, and other radioactivity in ground water, pp 193–203. CRC Press

Zarnke AM, Tharmalingam S, Boreham DR, Brooks AL (2019) BEIR VI radon: the rest of the story. Chem Biol Interact 301:81–87

Zeeb H, Shannoun F, World Health Organization (2009) WHO handbook on indoor radon: a public health perspective. World Health Organization

Chapter 3
Radon Distribution in Groundwater and River Water

Understanding the occurrence and factors controlling radon distribution in groundwater as well as surface water is important for improving its tracer applications in hydrology. The following sections and sub-sections explain the mechanisms influencing radon distribution, migration, and abundance in water.

3.1 Radon Distribution in Groundwater

Radon (^{222}Rn) is a colorless, odorless, chemically occurring radioactive isotope produced within uranium decay series, with radium as the penultimate parent. This radioactive ($t1/2 = 3.825$ days) alpha particle emitting gas is present ubiquitously in the rocks and soils of the earth's crust. Studies on radon are carried out worldwide from two perspectives. Studies on radon are carried out worldwide from two perspectives.

On one hand, ^{222}Rn has been studied from a health point of view, i.e., the radiation dose and carcinogenic risk due to radon. This aspect has been elaborated in Chap. 7. It is possible that radon is the second most important lung cancer-causing factor, after smoking (WHO 2009; UNSCEAR 2000), because more than 50% of natural radiation doses to the general public are delivered by radon, thoron, and their daughter products (UNSCEAR 2000, 2009; Robertson et al. 2013).

Another important aspect is the distribution of radon in environment (soil, water, and air) and its use in identifying and predicting earthquakes, volcanic activities, fault dislocation, and numerous applications in hydrological research (Terray et al. 2020; Sukanya et al. 2021; Frei and Gilfedder 2021; Manisa et al. 2022; Galiana-Merino et al. 2022; Schaper et al. 2022). Being a naturally occurring radionuclide, chemically inert gas, heavier than air, and emanating from rocks and soils, radon is ubiquitous in the environment and tends to accumulate in underground spaces, such as caves. Although radium occurs ubiquitously in all rocks and soils, its concentration varies with location and geological settings (Ravikumar et al. 2014). High radon

S. S. and S. Joseph, *Environmental Radon*, Environmental Science and Engineering,
https://doi.org/10.1007/978-981-99-2672-5_3

activity concentrations could be expected in uranium-mineralized regions. Several attempts have been made to correlate radon activity concentrations with factors such as geology, soil porosity, shears, thrusts, and faults (Heriawan et al. 2021; Sukanya et al. 2021; Arabi et al. 2021).

Unlike its immediate parent ^{226}Ra, which is a highly diffusive gas with free mobility, ^{222}Rn can typically flow through fractures, joints, pore spaces, and voids with ease, traveling considerable distances from its source of origin (Sukanya et al. 2021). The distribution of ^{222}Rn in groundwater is also influenced by factors like the degree to which various groundwater constituents mix in the circulation system, the residence time of groundwater, the distance or length of radon transport to reach the surface, etc. (Przylibski 2011).

The majority of the ^{222}Rn produced within a host mineral grain in parent rocks containing uranium and radium is retained within the grain, and only a little portion escapes to enter the pore space and reach the surface (Cinelli et al. 2019; Appleton et al. 2020). This fraction is called the emanation coefficient (K_{em}), i.e., the ratio between the amount of ^{222}Rn entering the pore space and the amount of radon generated from ^{226}Ra.

It should be noted that the spatial distribution of radium has a significant role in determining the emanation coefficient (Przylibski 2015). If the nuclei of ^{226}Ra isotope are found in the periphery of mineral crystals or are present in minerals nearby the surface of the cracks and fissures or pores, they release much more ^{222}Rn atoms into the pore space (and fissures) (Przylibski 2011, 2015; Przylibski and Gorecka 2014).

Radon dynamics have been the focus of extensive radon research in recent decades. The literature has identified several mechanisms by which radon moves, including alpha particle recoil from the grain surface, diffusion, and transportation through water or air (Giletti and Kulp 1955; Fleischer and Mogro-Campero 1978; Wood et al. 2004; Sakoda and Ishimori 2014).

The displacement of radon from mineral crystal lattices occurs due to radiation damage and results in emanation of alpha particles, i.e., alpha recoil with 4.78 meV of energy and 90 keV of ^{222}Rn atom (Pommé 2015). The migration of radon through micro-fractures and fissures facilitates its dissolution in water. The distance transported by radon through alpha recoil, known as the "alpha recoil length", can vary greatly depending on the host medium, ranging from 10^{-6} to 10^{-3} cm (Tanner 1964; Nguyen et al. 2018). The diffusion length, i.e., the distance traveled by radon through diffusion, can also vary based on the specific medium.

A shorter diffusion length in groundwater (a few centimeters) suggests that the groundwater radon is more likely dominated by transportation process, due to its short half-life (Gilfedder et al. 2015). Convection (i.e., vertical movement) and advection (i.e., lateral movement induced by pressure gradient) are two mechanisms controlling ^{222}Rn migration (Bonczyk et al. 2019). The solubility rate of ^{222}Rn depends on the nature of solution and its temperature (Ye et al. 2019).

Numerous studies have shown that ^{222}Rn emanation substantially correlates with basement geology, while ^{222}Rn contamination prevails in wells drilled in uranium-rich bedrocks (Cho and Choo 2019; Martins et al. 2019; Sukanya et al. 2021). There are multiple accounts of the presence of ^{222}Rn in aquifers around the world with various lithological motifs, including carbonate aquifers (Tallini et al. 2013), meta-sedimentary aquifers (Cho 2017), sedimentary aquifers (Cinti et al. 2019), basaltic aquifers (Ogden et al. 2020), crystalline aquifers (Przylibski et al. 2020), and alluvial aquifers (Sharma et al. 2020).

The variability in ^{222}Rn activity across different geological units, as well as within the same unit between neighboring wells, is attributed to several factors. These include geologic and structural characteristics, viz. rock type, emanation coefficient, and the presence of lineaments, as well as the length of the radon transport pathway and the residence time of groundwater (Przylibski and Gorecka 2014). The movement of groundwater within a basin is influenced by its topography and the underlying geological, structural, and topographic settings, which ultimately impact the rate of flow.

The migration of dissolved radon is restricted to the order of a few centimeters or meters a day, due to the relatively lower velocity of groundwater flow. After 40 days (~10 half-lives), only 0.1% of the ^{222}Rn atoms will exist. Hence, radon activity decreases with longer residence times. Therefore, ^{222}Rn activity in groundwater of a well is chiefly controlled by ^{226}Ra content, emanation coefficient of the reservoir rock (within a circle of a few to a few dozen meters radius), and lateral variation of geological formations (Durrani and Ilic 1997).

Mixing the proportion of various groundwater components within the circulation system is another determinant of ^{222}Rn activities in water (Przylibski 2011). For example, during the mixing process, deep circulation low radon waters get enriched with 222Rn when mixed with high radon shallow circulation waters (Przylibski 2011, 2018a, b; Przylibski et al. 2022) and this component is usually the infiltration fraction (Przylibski 2011; Belgacem et al. 2015).

Several rock types associated with high ^{222}Rn activities have been reported. Highly deformed and sheared uraniferous meta-sediments, volcanics, granite intrusives, glacial-lake-clay deposits having large surface area and high permeability (due to desiccation cracking), and marine black shales are the geologic domains with high activities of radon in groundwater have been observed in the United States (Gundersen et al. 1992). Studies by Scheib et al. (2013a, b) in England region noted high radon levels in groundwater in association with granites, breccias, and conglomerates of the Triassic, Mercia Mudstone Group marginal facies, shales, and cherts.

In India, multiple studies have documented the presence of ^{222}Rn in groundwater across various geological formations, including hard rock terrains (Thivya et al. 2015), alluvial aquifers (Sharma et al. 2020), and carbonate aquifers (Kumar et al. 2018). Radon studies have been conducted in groundwater in a range of lithological settings, such as the Chitradurga group, Bababudan group, Charnockites, Closepet granite, Migmatites and dolerites, and Satyamangalam schist complex (Ravikumar et al. 2014), the Garhwal group (composed of quartzite and metabasites) with intercalations of phyllite and chlorite schists, and the Central Crystalline group consisting of sheared granitic gneisses, porphyritic gneiss, talc schist, mica schist, mylonites, quartzo-feldspathic schist (Bourai et al. 2016), and Peninsular gneiss (Niranjan et al. 2017).

3.2 Case Study: Radon in Groundwater of Karamana River Basin, Southern Western Ghats, India

The occurrence of high ^{222}Rn in groundwater has been extensively investigated (Kalip et al. 2018; Naskar et al. 2018), but few studies have explored the underlying mechanisms explaining the variations in ^{222}Rn in groundwater (Przylibski 2011; Cho et al. 2019), particularly in tropical regions such as southwest India. Thus, Sukanya et al. (2021) conducted a study to understand the distribution and occurrence of ^{222}Rn in groundwater and identify the factors contributing to its abundance in the Karamana River Basin (KRB), a tropical mountainous river basin located in southwest India.

The KRB is a small basin covering an area of 702 km^2 situated between 8°22′–8°41′ N latitudes and 76°50′–77°14′ E longitudes in Kerala State, southwest India (Fig. 3.1). The basin comprises three physiographic divisions—highland (>75.0 m, amsl), midland (7.5–75.0 m, amsl), and lowland (<7.5 m, amsl)—that are characteristic of Kerala. The Karamana River, which is a sixth-order stream, originates at Chemmunji Motta peak (1717 m amsl) in the Western Ghats and flows westward through the highland–midland–lowland tracts before joining the Arabian Sea at Poonthura, southwest of Thiruvananthapuram City. The Peppara reservoir in the highland and the Aruvikkara reservoir in the midland provide drinking water to Thiruvananthapuram City. The Killi River, a major tributary of the Karamana River, originates from Nedumangad (8°40′ N, 77°00′ E) and joins the Karamana River at Kalladimukham (8°28′ N, 76°57′ E), approximately 4 km from the Arabian Sea.

With interludes of dolerite dykes, pegmatic, and quartzite veins, the KRB's geology displays an assemblage of Archaean crystalline rocks to modern alluvium. While khondalite, a garnet–biotite–sillimanite gneiss with or without graphite, underlies a large portion of the KRB, the remaining portions are also underlain by garnet–biotite gneiss with associated migmatites, quartzo-feldspathic–hypersthene granulite, and garnetiferous granite gneiss. Moreover, there are several places where basic (dolerite and gabbro) and acidic (pegmatites and quartz vein) rocks have been incorporated. A few spots of quartzo-feldspathic rocks and calc-granulites were found in KRB, according to Soman (1987).

The Karamana River flows through a region characterized by a uniform crystalline basement consisting mainly of garnetiferous quartzo-feldspathic gneisses of the khondalite suite. As a result of the nearly uniform composition of the bedrock, the river's channel network reflects more of the structural control than lithologic control. The channels have cut grooves into the gneiss and planation surfaces, and they follow the northwest–southeast axial traces of large-scale folds of gneissic foliation and banding that are almost upright (Sukanya et al. 2021).

Although the river's course is largely determined by the underlying structure, changes in the channel orientation and course are also influenced by conspicuous fracture zones and joint systems that are present at many locations. For example, at Aruvikkara, a well-established north–south trending vertical joint system regulates the river's north–south path. The river channel network is primarily influenced by the underlying structural control, which is mainly determined by the uniform crystalline

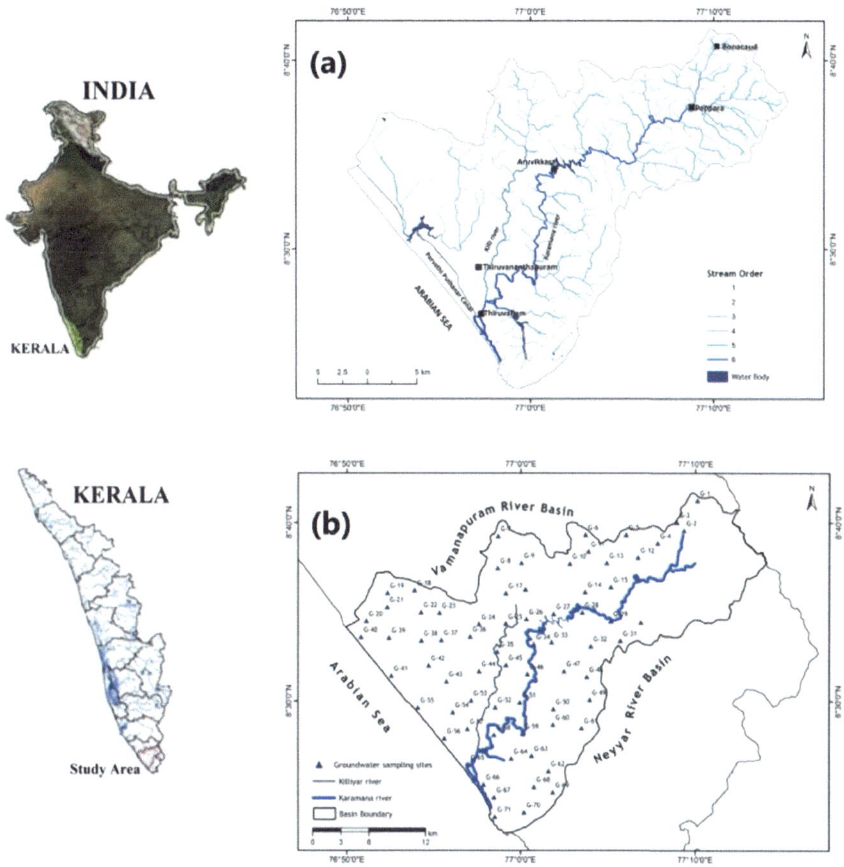

Fig. 3.1 Location map of the Karamana River Basin (KRB) showing **a** the drainage pattern **b** the sampling points (*Source* Sukanya et al. 2021)

basement (Sukanya et al. 2021). Nonetheless, fracture zones and joint systems also play a role in regulating the flowpath and channel orientation at various locations.

Despite the structural control on the channel network of river, frequent meanders are observed, especially in the midland and lowland tracts. The general trend of the river channel shows several shifts, with most of the changes occurring at the confluence of fourth-order tributaries, indicating a likely river capture rather than structural control. This suggests that the Karamana drainage system is complex and comprises both consequent and subsequent stream networks, as noted by Sinha Roy in 1979.

3.2.1 Water Sampling and Analytical Methods

In the study area, a total of 71 shallow dug wells, with a maximum depth of 25 m, were identified in a grid pattern at intervals of approximately 3 km (as shown in Fig. 3.1). The selection of the wells took into consideration the topography, geology, type (domestic/public well), accessibility, and condition of the wells. However, water sampling from the northeastern part of the study area was impeded by thick forest vegetation and low or lack of habitation.

Water samples were collected from the identified dug wells, and various physicochemical parameters such as water temperature, electrical conductivity (EC), and pH were measured at the site using a Horiba LAqua portable kit. Alkalinity was determined by titrating 10 ml of water sample with 0.02 N H_2SO_4, using methyl orange as an indicator. The alkalinity values are reported in milligrams of $CaCO_3$ per liter of water.

To obtain measurements of ^{222}Rn in groundwater, water samples were pumped out from the bottom of the dug well with a low-discharge peristaltic pump, and great care was taken to prevent contact with atmospheric air as the samples were collected in 250-mL glass bottles. Once in the laboratory, the dissolved ^{222}Rn in the water was stripped by circulating it through a closed air-loop via a desiccant tube and bubbling air through the water sample. Finally, the extracted ^{222}Rn was counted using a RAD7 ^{222}Rn counting system (Durridge make).

3.2.2 Radon Distribution

Table 3.3 in the Appendix presents the measurements of ^{222}Rn activities in groundwater from 71 locations in KRB during both non-monsoon (NON) and monsoon (MON) seasons in 2017 (Table 3.1).

The statistical data obtained from the study revealed that the standard deviation was greater than the median and arithmetic mean, indicating a highly variable non-normal distribution. The positive leptokurtosis (>3) and skewness also indicated the asymmetric distribution of radon, which could be attributed to the

Table 3.1 Statistical summary of ^{222}Rn activity (Bq/m^3) in groundwater samples, KRB (Data from Sukanya et al. 2021)

Statistical parameter	^{222}Rn activity (Bq/m^3)	
	Non-monsoon	Monsoon
Minimum	167	67
Maximum	68,352	34,595
Mean	7154.6	3230.6
Kurtosis	14.8	15.4
Skewness	3.6	3.6
Std. deviation	11,664.6	5559.9

geological and structural variations in the study area. The radon activity ranged from 167 to 68,352 Bq/m^3 during non-monsoon and from 67 to 34,595 Bq/m^3 during monsoon, with mean values of 7154.6 Bq/m^3 and 3230.6 Bq/m^3, respectively, showing significant spatial and seasonal variations.

Despite the altitudinal variations, the non-monsoon season showed high ^{222}Rn activities in isolated pockets of highland, midland, and lowland tracts (except coastal areas) of KRB. The spatial distribution contour map of ^{222}Rn in groundwater (Figs. 3.4 and 3.5) indicated that during non-monsoon, high ^{222}Rn activities (>11,000 Bq/m^3) were observed at eight locations, including G-1, G-27, G-42, G-43, G-52, G-62, G-63, and G-70, respectively. Among these, G-1 is in highland, G-27, G-42, G-43, G-52, G-62, and G-63 are in the midland, and G-70 is in the lowland region.

During monsoon, high ^{222}Rn activities (>11,000 Bq/m^3) were observed only at five locations, including G-1, G-27, G-62, G-63, and G-70, with radon activities declining significantly due to dilution. A large variability in ^{222}Rn activities was observed in groundwater, which is solely geogenic in origin. Therefore, to understand the relationship between radon anomaly and physiographic, geologic, and hydrogeologic aspects such as depth to water table, temperature, electrical conductivity, local geology/lithology, and structural features of the study area, further investigation is needed.

3.2.3 Frequency Distribution of ^{222}Rn in Groundwater

The results of the study on groundwater samples collected from 71 sites in KRB during the non-monsoon season revealed a significant striking range of variation in ^{222}Rn activities from 170 to 68,350 Bq/m^3. The frequency distribution pattern of the ^{222}Rn activities in all the groundwater samples is presented in Figs. 3.2, 3.3, where it was observed that the majority of samples (29.6%) showed activity levels within the range of 0–2,000 Bq/m^3, and the second highest frequency group (26.7%) had ^{222}Rn activity levels between 2,000 and 4,000 Bq/m^3. Surprisingly, the remaining eight sets of the least frequent groups (each covering 1%) exhibited remarkably high levels of ^{222}Rn activity, falling within the range of 14,000–70,000 Bq/m^3. These findings indicate that there is a significant variation in the ^{222}Rn activities in groundwater samples collected from different sites in KRB suggesting suggests the presence of diverse and complex factors influencing the spatial distribution of radon in groundwater. This highlights the need for further investigation to better understand those underlying factors contributing to such large variations.

During monsoon, radon activity levels exhibit significant variability, ranging from 67 to 34,595 Bq/m^3. The largest number of samples (60.2%) was within the range 0–2000 Bq/m^3, with the second largest group (5.8%) levels between 2000 and 4000 Bq/m^3. 5.6% of samples follow the sequence within the class 4000–8000 Bq/m^3.

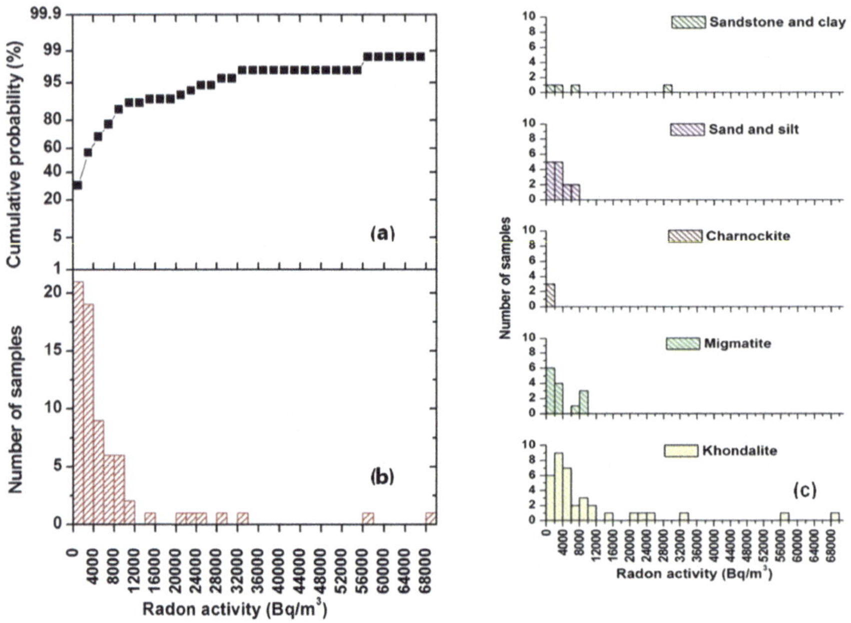

Fig. 3.2 Frequency distribution of ^{222}Rn in groundwater, **a** cumulative probability, **b** all samples, **c** in various rock formations, KRB during NON-2017 (*Source* Sukanya et al. 2021)

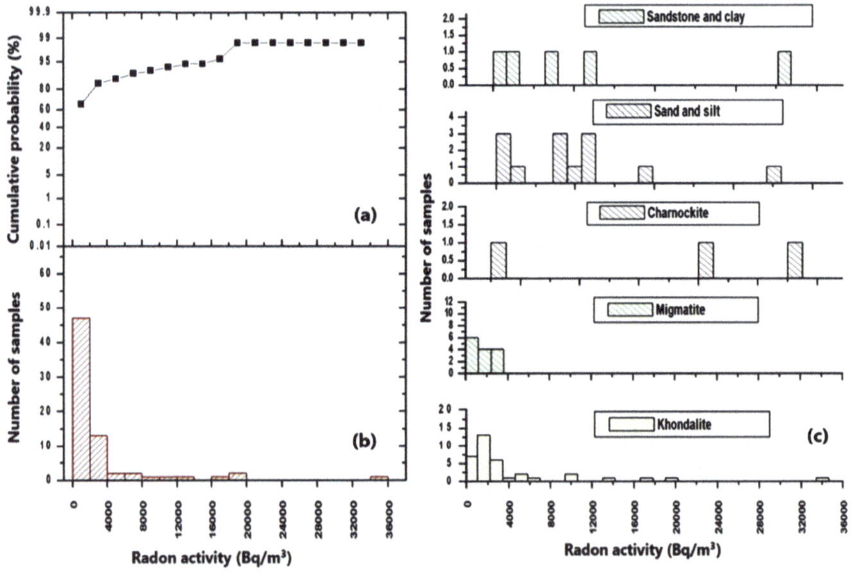

Fig. 3.3 Frequency distribution of ^{222}Rn in groundwater, **a** cumulative probability, **b** all samples, **c** in various rock formations, KRB during MON-2017

A smaller frequency group covering 2.8% fall under the range 18,000–20000 Bq/m^3. The least number of samples (1% each) shows ^{222}Rn activity within two ranges, i.e., 8000–14,000 and 34,000–36,000 Bq/m^3.

The gneissic aquifers in KRB have ^{222}Rn activity levels that fall within the range of a global dataset compiled by Girault et al. (2018), which is from 7.38 $\times$ 10^{-4} to 10,2000 Bq/L, during both non-monsoon (=170–68,350 Bq/m^{3,} i.e., 0.17–68.35 Bq/L) and monsoon (=92–34,595 Bq/m^{3,} i.e., 0.09–34.59 Bq/L) seasons.

However, the activity levels in KRB tend to be towards the lower range of this dataset. The groundwater in KRB can be categorized as radon-free water (>1000 Bq/m^3), radon-poor water (1000–9,900 Bq/m^3), and low radon water (10,000–99,900 Bq/m^3) based on the classification proposed by Przylibski (2005), indicating that the region has a relatively low radon concentration. These results suggest that KRB has a lower risk of radon exposure compared to areas with higher radon activity, which could have significant implications for public health and safety.

3.2.4 Spatio-Temporal Variability of ^{222}Rn in Groundwater

The spatial variability of ^{222}Rn activity in groundwater is influenced by several factors, including changes in geology, lithological framework, soil types, and structural controls. The spatial distribution contour map of ^{222}Rn in groundwater (Figs. 3.4, 3.5) revealed high ^{222}Rn activities in isolated pockets of highland, midland, and lowland tracts of KRB, regardless of altitudinal variations. Figure 3.5 also indicates relatively lower ^{222}Rn activities in wells located in the western part of KRB.

The decreasing trend of ^{222}Rn activity from the northeast towards the central regions of the basin and further towards the southwest coastal zone is attributed to the depth of bedrock and the residence time of groundwater. This trend corresponds to the general direction of groundwater flow (NE to SSW) in the shallow aquifers of KRB (Fig. 3.4). In the highland region (northeast part of the basin), the compact hard rock is exposed near the surface, and the thickness of weathered sediments is minimal (<5 m). The groundwater flow in this region is mainly through weathered and interconnected fractures, resulting in fast flow velocities (due to secondary porosity) and short residence time of groundwater compared to the sandy aquifer in the coastal areas.

The short residence time of groundwater in the highland region may explain the higher radon activity observed in this area. As radon gets dissolved along the flow path without appreciable radioactive decay, the dissolved radon concentration can increase before reaching the well. Hydraulic turnover time of groundwater is increasingly recognized as a significant factor controlling ^{222}Rn activity. Martins et al. (2019) concluded in their study that high-turnover time of springs leads to low groundwater radon activity in the absence of an anomalous radioactive source near the surface.

The thickness of quaternary alluvium increases in the coastal areas and the depth to bedrock increases towards the south–southwestern parts of the basin, leading to

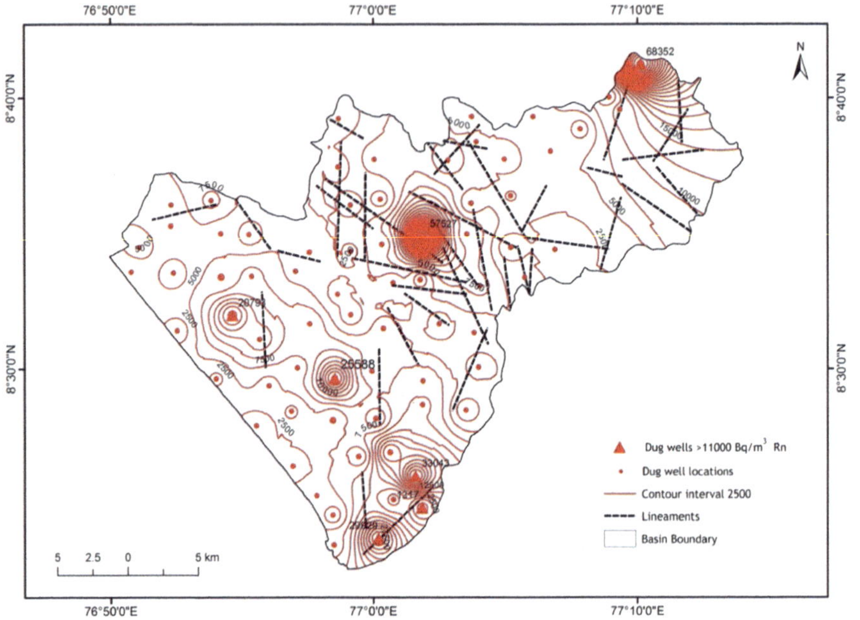

Fig. 3.4 Spatio-temporal variation of ^{222}Rn in groundwater, non-monsoon (*Source* Sukanya et al. 2021)

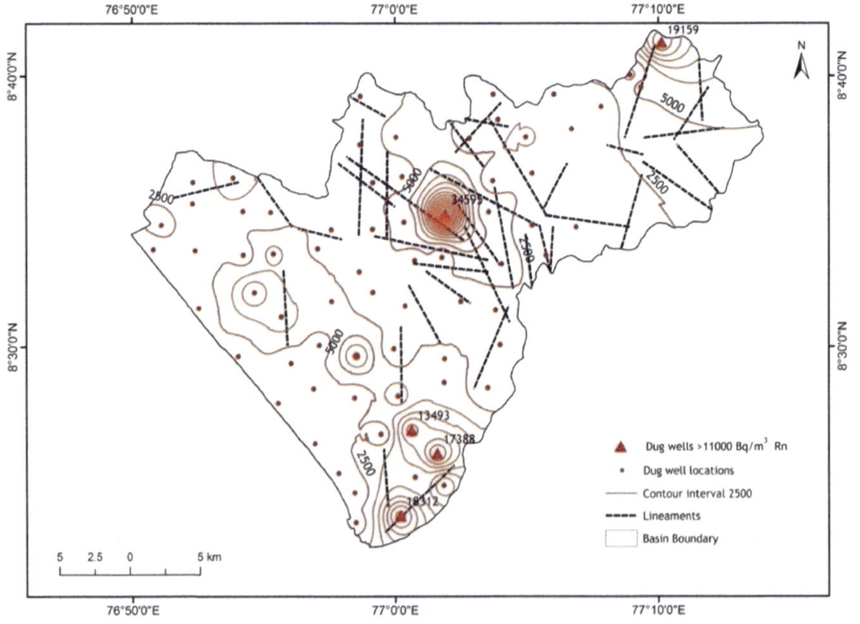

Fig. 3.5 Spatio-temporal variation of ^{222}Rn in groundwater, monsoon

a decrease in radon activity. In the coastal region, groundwater had relatively low ^{222}Rn activities, and no ^{222}Rn anomaly was observed. This could be attributed to the radioactive decay of radon, which occurs due to the slower flow velocities and longer residence times of groundwater in the coastal sandy aquifer.

The southwest coast of India is known for its high natural radioactivity, primarily due to the presence of the radioactive mineral monazite in the black sand placers that originate from the weathered parent rock and are transported by rivers flowing from the Western Ghats (Raju et al. 1986). Ramsiya et al. (2019) reported higher activities of ^{226}Ra, which is the immediate parent of ^{222}Rn in the coastal tract sediments of Chavara and Kollam in Kerala. In the western part of the study area, Lekshmi et al. (2018) found an average ^{226}Ra activity of 58.5 Bq/kg for the coastal sediments of Kovalam.

3.2.5 *Relationship of ^{222}Rn in Groundwater with Physical Parameters*

The shallow aquifers in KRB showed a wide range of depth to water table, varying from 1.6 m (at G-66 in the coastal region) to 25.6 m (at G-68 in the southeastern part) during non-monsoon and 1.1–18.63 m during monsoon. Despite this variability, there was no observable correlation between water table level and ^{222}Rn activity in groundwater. As seen in Fig. 3.6, the depth-wise distribution of ^{222}Rn in groundwater across different rock formations also showed no clear pattern. For example, the highest observed radon activity of 68,350 Bq/m^3 at location G-1 in the highland area had a relatively low depth to water table of 2.8 m.

The results of this study support previous findings that small-scale depth to water table does not have a significant impact on ^{222}Rn levels (Nandakumaran and Vinayachandran 2020). However, larger depth scales may show variation in ^{222}Rn activity when comparing shallow circulation waters at depths of 0–50 m with deep circulation waters at depths of several hundred meters to several kilometers (Singaraja et al. 2016).

According to research by UNSCEAR (2000), water temperature is one of the key factors that affect the degree of ^{222}Rn dissolution in water. This is due to the fact that ^{222}Rn is a diffusive gas, and its solubility in water decreases as temperature rises, causing it to escape more quickly. In the study area, high levels of ^{222}Rn were observed in groundwater with relatively lower temperatures (23.0–23.9 °C) in the khondalite terrain at G-1. However, when all samples were considered, only a weak negative correlation ($r^2 = 0.17$; $n = 36$ in non-monsoon, $r^2 = 0.23$; $n = 36$ in monsoon) was observed between ^{222}Rn activity and groundwater temperature in the khondalite terrain (Fig. 3.6).

A strong negative correlation ($r^2 = 0.65$ in non-monsoon, $r^2 = 0.69$ in monsoon; $n = 9$) was observed when only high ^{222}Rn groundwater (>11,000 Bq/m^3) was considered. However, in the case of charnockite rock, although ^{222}Rn activities were

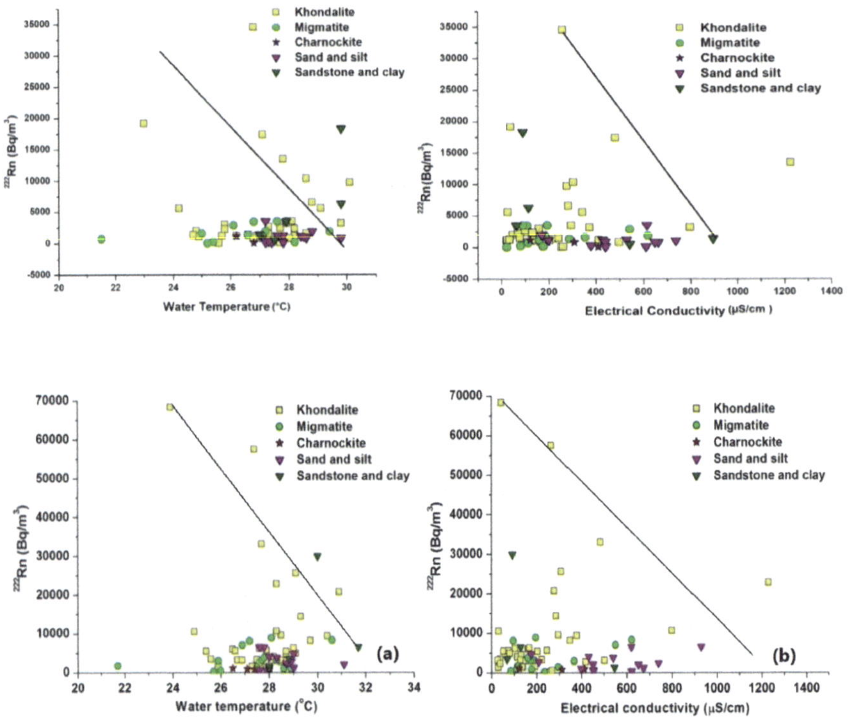

Fig. 3.6 Correlation of ^{222}Rn activity with water temperature and electrical conductivity w.r.t various rock formations, KRB (solid line implies the approximate trend). *Source* Sukanya et al. (2021)

lower (1200 Bq/m^3), a significant negative correlation ($r^2 = 0.98$ in non-monsoon, $r^2 = 0.86$ in monsoon; $n = 3$) was observed. Nevertheless, the dependency of ^{222}Rn activity is generally observed only when samples with temperature differences of 30–50 °C are compared (Tabar and Yakut 2014).

Previous studies have reported an inverse relationship between ^{222}Rn and EC of groundwater, indicating that low-mineralized waters of shallow circulation tend to have higher activity of ^{222}Rn (Przylibski 2011). In KRB, only the charnockite lithologic unit showed a significant negative correlation of EC with ^{222}Rn ($r^2 = 0.72$; $n = 3$) during non-monsoon. Some high ^{222}Rn activity sites in khondalitic (gneissic) formation also showed a slight negative correlation with EC (Fig. 3.6).

Therefore, it is evident that the large variability of ^{222}Rn activities in groundwater can be attributed to several geological and hydrogeological factors, which are evaluated and discussed in the subsequent sections.

3.2.6 Variation of ^{222}Rn in Groundwater with Various Rock Types

Geological factors, viz. rock type, macro-structure, and depth of weathering, have been identified as the most significant factors controlling the source and distribution of ^{222}Rn in groundwater (Scheib et al. 2013a, b; Watson et al. 2017). Near-surface weathering and bedrock fracturing have been suggested to contribute to the increase in ^{222}Rn activity in groundwater in various locations worldwide (Przylibski 2011).

Several studies conducted in different terrains have reported a direct correlation between the mineralization of uranium and thorium in the parent rock and the enrichment of ^{222}Rn levels in groundwater (Thivya et al. 2015). These findings suggest that geological factors play a crucial role in determining the concentration of ^{222}Rn in groundwater (Table 3.2).

The geological map of KRB reveals that the dominant rock type is khondalite, which is characterized by garnetiferous biotite–sillimanite gneiss. This rock type is intruded by migmatites, which trend in an NW–SE direction and indicate lineament-controlled emplacement (Fig. 3.4). However, in many areas of the highland and midland regions, the khondalite is overlain by thin to thick layers of lateritic rocks.

These gneissic rocks are typically compact but vulnerable to chemical weathering and contain the radioactive mineral monazite (Akingboye et al. 2021). They also contain trace amounts of radioactive decay products such as uranium, thorium, and radium. Weathering processes transport these decay products from the parent rock and precipitate or re-deposit them in favorable geochemical environments downstream, leading to the release of more ^{222}Rn into the aquifers.

In KRB, 50.7% of the selected dug wells for ^{222}Rn sampling are located in the khondalite terrain, followed by 19.7% each in sand and silt and migmatite aquifers, with sandstone with clay (5.6%) and charnockite (4.2%) formations following thereafter (Supplementary material 2). Of note, 75–80% of the elevated ^{222}Rn activities in groundwater were found in khondalite, with the remainder (20–25%) confined to sandstone and clay during both seasons. Furthermore, 68% of the 71 investigated

Table 3.2 Distribution of Wells and exceedance of ^{222}Rn activity above USEPA limit in different rock formations of KRB (Sukanya et al. 2021)

Rock type	No. of wells	Percentage of wells (%)	No. of wells with ^{222}Rn activity > 11,000 Bq/m^3	
			Non-monsoon	Monsoon
Khondalite	36	50.7	6	4
Sand and silt	14	19.7	–	
Charnockite	3	4.2	–	
Sandstone and clay	4	5.6	2	1
Migmatite	14	19.7	–	

wells had penetrated through a laterite cap, which is a ubiquitous end product of weathering of basement rocks in a tropical climate with distinct wet and dry spells.

Figure 3.7 illustrates the frequency distribution of ^{222}Rn in groundwater for each major geological rock type in KRB. The majority of groundwater samples are located in the khondalite terrain, showing a wide range of ^{222}Rn activity, ranging from 170 to 68,350 Bq/m^3 during non-monsoon and 67–34,595 Bq/m^3 during monsoon. The abundance of radon varies across different rock types, with the highest concentration found in khondalite and the lowest in sandstone and clay (Fig. 3.7).

Migmatite, which constitutes 19.7% of the rock types, has ^{222}Rn activity ranging from 170 to 8,880 Bq/m^3. Of the 14 samples in this rock type, 71.4% have ^{222}Rn activity within the range of 0–4,000 Bq/m^3, while 28.5% fall within the range of 6,000–10,000 Bq/m^3.

Sand and silt account for 19.7% of the total samples in KRB, with a significant range of ^{222}Rn activity observed from 320 to 6,440 Bq/m^3. Of the 14 samples in this category, 71.4% have ^{222}Rn activity within the range of 0–4,000 Bq/m^3, while 28.5% fall within the range of 4,000–8,000 Bq/m^3.

Sandstone and clay make up 5.6% of the rock types, with only 4 samples out of the total 71 samples. A wide range of ^{222}Rn activity is observed in this geologic unit, ranging from 1,320 to 29,930 Bq/m^3. Of this category, 50% fall within the range of 0–4,000 Bq/m^3, while 25% lie within the range of 6,000–8,000 Bq/m^3, and the remaining 25% fall within the range of 30,000–32,000 Bq/m^3.

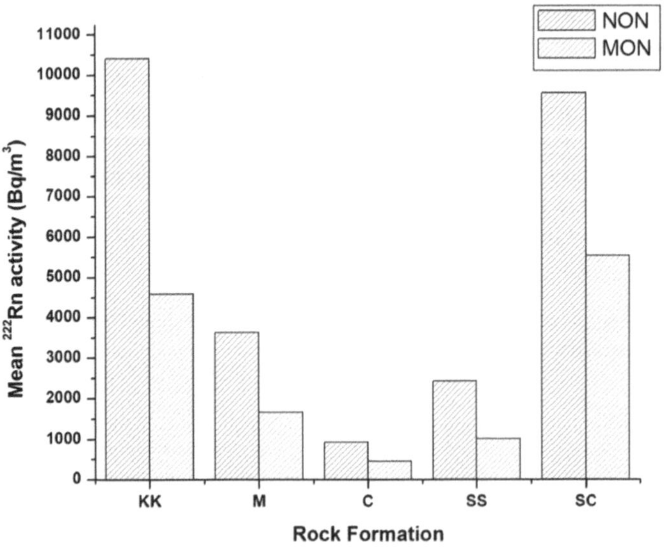

Fig. 3.7 Seasonal variation in groundwater ^{222}Rn in various rock formations; KK-Khondalite, M-Migmatite, C-Charnockite, SS-Sand and Silt, SC-Sandstone and Clay

Charnockite is a minor rock type in KRB, covering only 4.2% of the total number of samples, with a narrow range of ^{222}Rn activity from 800 to 1,110 Bq/m^3. All the samples in this category fall within the range of 0–2,000 Bq/m^3.

The khondalitic (Gneissic) aquifer in KRB exhibits the highest ^{222}Rn activity among the aquifers studied, with a maximum value of 68,350 Bq/m^3. This value is greater than the reported activities in the Varahi and Tuticorin South Indian hornblende biotite gneiss aquifers (10,100 and 40,700 Bq/m^3, respectively) by Somashekar and Ravikumar (2010) and Singaraja et al. (2016), respectively. In Sudan, Idriss et al. (2011) found a wide range of ^{222}Rn activity (1,580–345,100 Bq/m^3) in acid-gray gneissic and granitic aquifers.

A review by Girault et al. (2018) on ^{222}Rn analysis of groundwater samples from different continents showed that the ^{222}Rn activity varies widely with respect to different rock types. Basaltic and tidal aquifers exhibit low ^{222}Rn activity (30–15,000 Bq/m^3), while sediment and meta-sediment aquifers show slightly higher activity (2,000–3,000,000 Bq/m^3). Granitic aquifers exhibit higher activity (15,000–100,000,000 Bq/m^3), and deep Stripa granite aquifers (300–1200 m) exhibit the highest activity (above 100,000,000 Bq/m^3). These findings highlight the significant influence of bedrock composition on ^{222}Rn activity in groundwater.

In order to further understand the behavior of ^{222}Rn based on variations in lithology, the lithostratigraphy of wells located at G-11, G-20, G-53, and G-62 in KRB was examined using data from CGWB (unpublished). The lithology at G-11 consisted of thick weathered garnetiferous gneiss underlain by alternating layers of massive and fractured gneiss. In contrast, G-20 had a thick layer of sand underlain by sand and clay layers. At G-53, a thick layer of clay was found to be underlain by fractured garnetiferous biotite gneiss, while G-62 had a thick layer of laterite above fractured khondalite. The formation of mineralizing fluids enriched with uranium can be associated with higher grades of metamorphism (Mercadier et al. 2013), which may explain the high uranium levels observed in garnet biotite gneiss and garnet sillimanite gneiss rock types of khondalite group in Kerala Khondalite Belt (KKB) (Ray et al. 2008). These results suggest that bedrock uranium plays a significant role in the high levels of ^{222}Rn observed in gneiss aquifers of KRB.

Ray et al. (2008) conducted an analysis of uranium variation in the garnet biotite gneiss and garnet sillimanite gneiss rock types of the khondalite group in the Kerala Khondalite Belt (KKB), which showed very high uranium levels (max = 12.4 ppm; mean = 3 ppm) in garnet biotite gneiss and very low levels in charnockites (mean = 0.9 ppm). Meanwhile, Braun (2006) reported slightly lower uranium levels (mean = 2.6 ppm) in granitic gneiss of KKB. These results are consistent with the high radon levels observed in the gneiss aquifers of KRB, indicating that bedrock uranium is an important source of radon.

Furthermore, monsoonal variation of ^{222}Rn with respect to various geological formations (as depicted in Fig. 3.8) was analyzed. The results indicated that the sand and silt category exhibited the highest degree of variation in ^{222}Rn level, with a decrease of 58.5% during monsoon. Conversely, sandstone and clay formations showed the least variation (42%) as a result of monsoon rainfall.

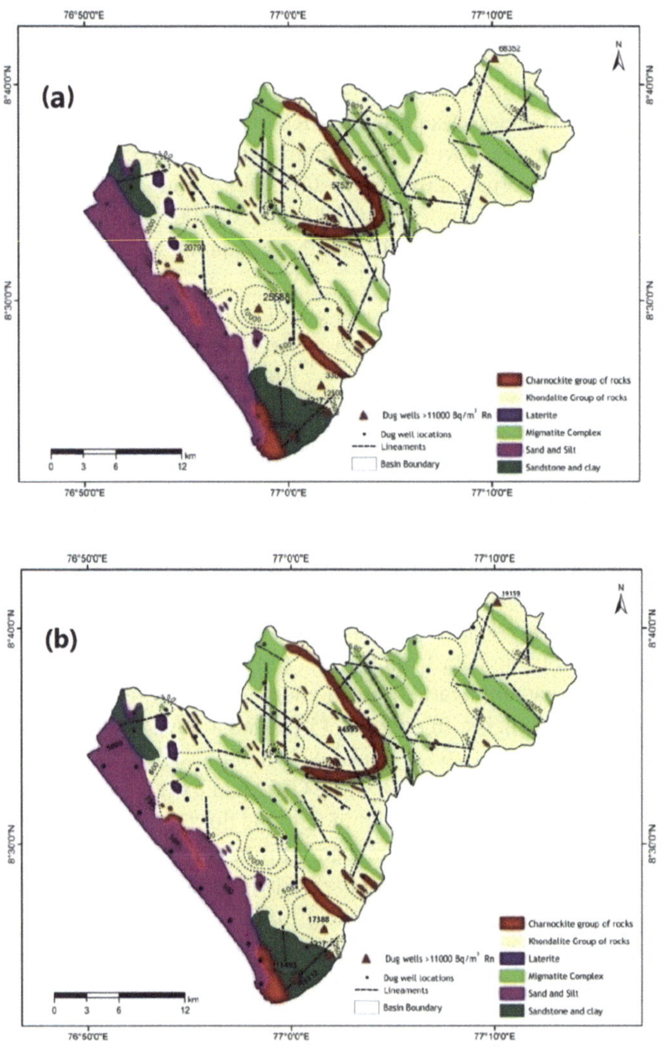

Fig. 3.8 Spatio-temporal variation of ^{222}Rn versus geology of the area (*Source* Sukanya et al. 2021)

3.2.7 Relationship of ^{222}Rn in Groundwater with the Emanation Coefficient

The high activity of ^{222}Rn in low mineralized groundwater with shallow circulation (within a few dozen meters of the ground level) through radium-rich reservoir rock is primarily due to the greater emanation coefficient in shallow depths, which is caused by intense weathering, crack density, and empty spaces in an aquifer (Przylibski 2011).

The dissolved activity of ^{222}Rn in groundwater (cpRn, in Bq/m^3) depends on several factors, including the activity of its parent ^{226}Ra in the reservoir rock (q, in Bq/kg), the emanation coefficient of the rock (Kem), the physical properties of the rock that influence its permeability to water and gases, such as the grain density (ρs, in kg/m^3) and the effective porosity (n), and the activity of $^{226}Ra2+$ ions dissolved in the groundwater (cpRa0, in Bq/m^3). The mathematical relationship between these factors is given by Przylibski (2011)

$$c_{pRn} = \frac{(1-n)\cdot\rho_s\cdot q\cdot K_{em}}{n} + c_{pRa0} \tag{3.1}$$

The emanation coefficient typically decreases as the depth of aquifers increases in lithologies with simple structure. Highly weathered, fractured crystalline aquifers, composed primarily of gneiss and granite, exhibit high emanation rates (Przylibski 2018a, b).

This study evaluated the emanation coefficient of rock using Eq. (3.1) with average geological characteristics for the highland, midland, and lowland regions. The parameters used in the equation are as follows:

- For the fractured granitic aquifer in the highland region: cpRn = 68,350 Bq/m^3 (the maximum observed activity in the area); q = 154 Bq/kg (calculated based on the maximum uranium content of 12.4 ppm in garnet biotite gneiss reported by Ray et al. 2008, assuming secular equilibrium between the activities of 238U and ^{226}Ra in the rock); $n = 0.15$; ρs = 2500 kg/m^3; cpRa0 = 35 Bq/m^3 (as reported by Nagabhushana et al. 2020 in a nearby basin with similar geology).
- For the weathered lateritic aquifer (midland)—cpRn—30,000 Bq/m^3 (the average activity in this area); q—32.3 Bq/kg (calculated based on the average uranium content of 2.6 ppm in the granitic gneiss of KKB as reported by Braun (2006)); n—0.25 (Bhosale and Kumar 2002); ρs—2500 kg/m^3; cpRa0—35 Bq/m^3 (Nagabhushana et al. 2020).
- For the sandy alluvial aquifer (lowland)—cpRn—2,500 Bq/m^3 (the average activity in this area); q—58.5 Bq/kg (as reported by Lekshmi et al. (2018) for the western part of KKB); n—0.35 (Todd and Mays 2005); ρs—2500 kg/m^3; cpRa0—35 Bq/m^3 (Nagabhushana et al. 2020).

The research findings suggest that the emanation coefficient tends to decrease with greater depth in aquifers with simple lithology. In contrast, fractured crystalline aquifers consisting mainly of gneiss and granites with high weathering rates exhibit high emanation rates of radon (Przylibski 2018a, b). The emanation coefficient, greater in shallow depths due to intense weathering, crack density, and empty spaces in an aquifer, causes high ^{222}Rn activity in low mineralized groundwater with shallow circulation through the radium-rich reservoir rock (Przylibski 2011).

The calculated values of the emanation coefficient for the three aquifers in the KKB region showed significant variations, with the highest value observed in the fractured granitic aquifer (highland) and the lowest in the sandy alluvial aquifer (lowland). The representative values of the emanation coefficient for the fractured granitic aquifer, weathered lateritic aquifer (midland), and sandy alluvial aquifer (lowland) were found to be 0.03, 0.13, and 0.009, respectively. These values fall within the normal range of 0.01–0.1 as reported in previous studies (Przylibski 2011).

The low emanation coefficient in the sandy alluvial aquifer (lowland) suggests that a majority of the ^{222}Rn produced from the decay of ^{226}Ra in the host rock was not released into the groundwater, but instead retained in the rock grains. This explains the observed low activity of ^{222}Rn in this aquifer. In contrast, the high emanation coefficient observed in the fractured granitic aquifer (highland) indicates that a significant amount of ^{222}Rn was released into the groundwater due to the intense weathering, cracks, and empty spaces in the aquifer. The weathered lateritic aquifer (midland) exhibited an intermediate value of emanation coefficient, which was consistent with the lithological characteristics of the aquifer.

The similarity in the low value of emanation coefficient between the present study and the study by Pinto et al. (2020) in the Chavara area indicates the importance of geological factors in determining the emanation coefficient. The high value of emanation coefficient observed in the weathered lateritic aquifer (midland) indicates the presence of a dense crack network and/or high porosity in the reservoir rock, which could be attributed to the intensive weathering and tectonic deformations in the region.

These findings are consistent with the observations made by Przylibski (2011) on the influence of reservoir rock characteristics on the emanation coefficient. The parameters such as ^{238}U/^{226}Ra concentration in rocks and emanation coefficient are less influenced by climatic and anthropogenic variables compared to ^{222}Rn activity in groundwater, soil, and air environments, as reported by Pereira et al. (2017).

3.2.8 Influence of Structural Features on ^{222}Rn Variability in Groundwater

While geological structures may not always have a significant impact on radon transport, tectonically active regions often exhibit elevated levels of ^{222}Rn in groundwater resulting from the release of radon from deeper sources. In areas with high tectonic activity, such as KRB, fractures, and faults can facilitate the movement of radon from greater depths to the surface, resulting in high ^{222}Rn activities in groundwater and soil (Ioannides et al. 2003).

The dominant rock type in KRB is khondalite, which is a garnetiferous biotite–sillimanite gneiss. Despite having low primary porosity, these rocks can become repositories for groundwater due to the development of secondary porosity resulting from lineament-aligned fracturing and chemical weathering. The rate of groundwater storage and yield is primarily dependent on the characteristics of the fractures and the rate of infiltration through them. Deformation of these rocks due to tensional or shear stresses can also result in the formation of fractures, which in turn can increase the infiltration rate and radon emanation. Furthermore, chemical weathering can facilitate the precipitation of ^{226}Ra atoms in pores and fissures, leading to an increase in ^{222}Rn emanation (Przylibski 2011).

KRB is a tectonically disturbed terrain with a dense set of lineaments that serve as conduits for rainwater seepage and radon gas transport. The lineament map of the KRB shows that most of the lineaments are concentrated in the central and northeastern part of the basin, with different orientations and these lineaments occasionally cross-cut each other at certain locations (Fig. 3.8). One major set of lineaments trends north–south, while another set trends northwest–southeast, and a less prominent set aligns along northeast–southwest.

Lineaments that are related to tectonic dislocation zones or zones with intense geochemical changes facilitate radon transport to a significant extent. In certain areas of the KRB, relatively higher radon activities have been observed near lineaments that are linked to intense weathering characterized by thick lateritic cappings on top of crystalline rocks. Moreover, tectonic deformations in these regions amplify the emanation coefficient of rocks, which leads to high ^{222}Rn activity. This is in line with the findings of Przylibski (2011).

The presence of both geologic and lineament factors contributes to the high ^{222}Rn anomalies detected at certain locations within the KRB. For instance, location G-1, situated in the highland region of the northeastern part, exhibits the highest ^{222}Rn anomaly (68,350 Bq/m^3) and is situated in a terrain of khondalite group of Precambrian high-grade metamorphic rocks (Chacko et al. 1987). According to Soman (2002), most of the crystalline rocks in this region were subjected to polymetamorphic and poly-deformational events. In addition, Knutsson and Olofsson (2002) suggest that high ^{222}Rn areas are often found in regions lacking bedrock of younger granites. Furthermore, it has been observed that areas with higher erosion display lower ^{222}Rn activity, while regions with lesser erosion are associated with higher ^{222}Rn activity (Przylibski and Gorecka 2014; Przylibski et al. 2020).

The high ^{222}Rn content observed at location G-1 is likely due to the combination of two factors: (a) the presence of acidic pegmatites, which are known to be rich in uranium, leading to increased uranium enrichment and the formation of relatively high radioactivity of daughter products, including ^{222}Rn; and (b) the proximity of the well to lineaments (Fig. 3.8), which likely act as a channel through which ^{222}Rn is transported via groundwater flow. Similarly, location G-27, situated in the midland region, recorded the second highest ^{222}Rn activity (57,530 Bq/m^3) and is also located in a khondalite rock terrain and a lineament-dense region enclosed within a charnockite semi-loop of a fold. The presence of high ^{222}Rn activity coincides with the location of lineaments, suggesting that this region is potentially suitable for groundwater development.

At location G-52 in the KRB midland region, the major rock type is garnetiferous biotite gneiss, overlain by clay. The elevated ^{222}Rn activity at this site is inferred to be caused by the foliations and fractures. The high alkalinity (360 mg/L) indicates active weathering, which may have contributed to the release of more ^{222}Rn from the rock into the groundwater.

The lithology at location G-63 in the midland region shows a surface layer of laterite overlying sandy bottom layers. The high permeability of sand favors lateral advection of ^{222}Rn. The nearby site G-62 has a laterite cap (11.5 m) overlying khondalite, and exhibits very high ^{222}Rn activity. The presence of fractures in most layers suggests that the fine-grained and weathered khondalites are easy to escape pathways for ^{222}Rn to dissolve into groundwater. Location G-70, a relatively deep well (depth = 19.3 m), also has high ^{222}Rn activity (29,930 Bq/m^3), and is located in the midland region. It penetrates through a laterite cap (depth = 10 m) and into khondalitic rocks. The high ^{222}Rn activity in this location may be attributed to the high uranium/radium content and emanation coefficient of the rock.

In the lowland region, wells G-71 and G-20 exhibited relatively high ^{222}Rn activity (6,570 Bq/m^3 and 6,440 Bq/m^3, respectively). The presence of weathered garnet biotite gneiss at location G-20 may have contributed to the higher ^{222}Rn activity in groundwater. However, wells in the beach sand in the coastal area showed relatively low ^{222}Rn activity.

In other areas, although the ^{222}Rn activity was generally low, wells located closer to lineaments exhibited relatively higher ^{222}Rn activity compared to wells located farther away. This observation suggests that most of the high ^{222}Rn anomalies are concentrated at or near major lineaments in KRB. Three radon anomalies (locations G-42, G-43, and G-52) in the midland occur along the SE–NW direction, coinciding with a lineament, indicating macro-structural control on the release of ^{222}Rn into the groundwater. These lineaments may be considered active lineaments. Notably, locations G-62 and G-70, which had elevated ^{222}Rn levels, lie along the same NE–SW trending lineament. Jacob et al. (2009) reported the association of high ^{222}Rn values with lineaments in Vizhinjam, the southern part of KRB.

Nevertheless, it is important to note that certain sites located close to lineaments do not exhibit high ^{222}Rn activity, which suggests that these lineaments may be tectonically inactive. Conversely, locations such as G-52, G-42, and G-43, which are not located near any lineament, demonstrate high ^{222}Rn activity, indicating the presence of gas-permeable buried fracture or fault zones that are still tectonically active. These zones may not be exposed at the surface and can be identified through satellite imagery. High uranium/radium content in the rock and a high emanation coefficient may also contribute to high ^{222}Rn levels in groundwater in these areas.

Based on Figs. 3.9 and 3.10, five major factors can be identified as controlling ^{222}Rn in the groundwater of KRB. However, previous research has demonstrated that including too many factors (>8) can result in multi-collinearity among variables, which can decrease the accuracy of prediction and modeling.

The ^{222}Rn activity in the groundwater of KRB is influenced by several factors, which can be classified into five major categories. The first factor is the *proximity to active lineaments*, where samples belonging to group I with high ^{222}Rn activities are located close to the lineaments at distances less than 100 m. The second factor is the *proximity to inactive lineaments*, where groundwater samples with relatively moderate to lower ^{222}Rn activities and located close to the lineaments (at distances less than 100 m) are classified as group II. The third factor is a *combination of* ^{226}*Ra content and emanation coefficient of the reservoir rock*, where group III samples with high ^{222}Rn activities are located far away from the lineaments (>100 m). Other driving factors such as *uranium mineralization in the aquifer and proximity to buried faults*

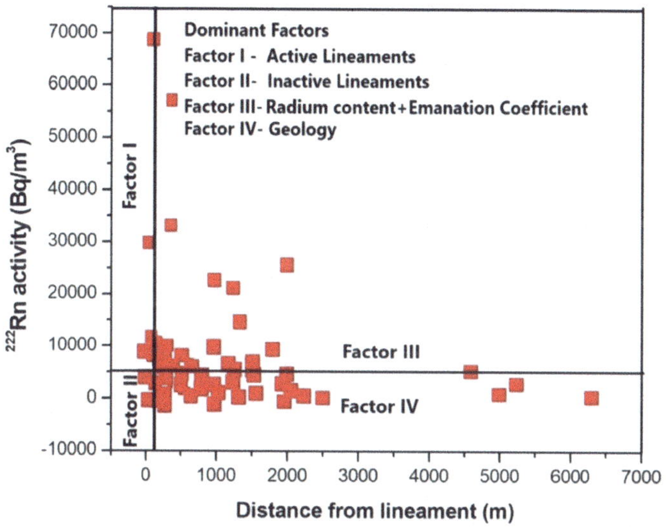

Fig. 3.9 Delineation of factors controlling ^{222}Rn activity in groundwater w.r.t the proximity to the lineaments in KRB, non-monsoon. (*Source* Sukanya et al. 2021)

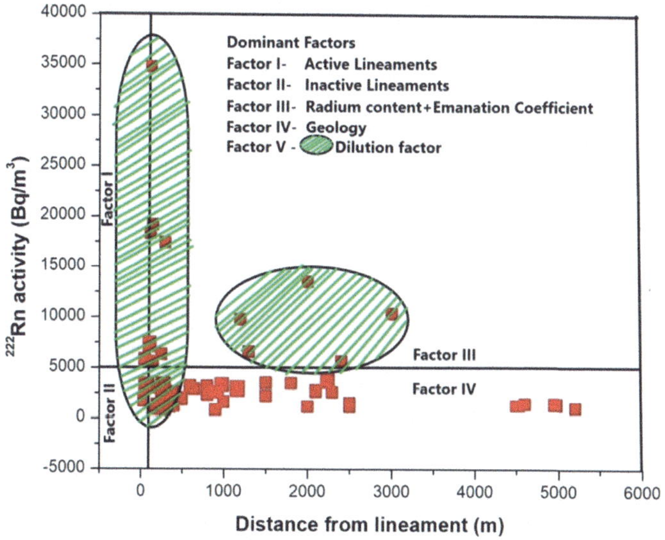

Fig. 3.10 Delineation of factors controlling ^{222}Rn activity in groundwater w.r.t proximity to the lineaments in KRB, Monsoon

could also be responsible for high ^{222}Rn activities. The fourth factor is the *influence of bedrock geology*, where group IV samples collected away from the lineaments have relatively moderate to lower ^{222}Rn activities due to the type and depth of the bedrock, residence time, and distance of transportation of radon dissolved water. The fifth factor is dilution due to precipitation, which results in a decrease of 2–4 times in groundwater radon during the monsoon period.

3.2.9 *Minor Factors Controlling Groundwater ^{222}Rn*

Apart from major factors, viz. uranium concentration and/or radium (^{226}Ra) activity of the bedrock and/or soil, as well as the radon emanation coefficient influencing groundwater radon (Gruber et al. 2013; Pereira et al. 2017; Sukanya et al. 2021), two minor factors were also investigated in the study.

(a) **Radon production potential**

Radon production potential/rate has been enlisted as one of the influencing factors in recent studies (Pereira et al. 2017; Martins et al. 2019). Radon production rate (P_{Rn}, in Bq m^{-3} h^{-1}) was also calculated using the following equation:

$$P_{Rn} = A_{Ra} E d \lambda \tag{3.2}$$

where A_{Ra}-^{226}Ra is the mass activity (Bq kg^{-1}), E is the emanation coefficient, and d is the apparent density, λ is the decay rate of ^{222}Rn, i.e., (ln2/(3.8 × 24 h) = 0.00761/h. [ln(2) = log$_e$(2) = 0.6931; λ = 0.6931/(3.8 × 24) = 0.00761].

In the present study, the radon production rate of the rock was evaluated using Eq. (3.2) for typical average geological units (highland, midland, and lowland). The parameters used in the equation include:

For the fractured granitic aquifer (highland)—A_{Ra}–154 Bq/kg (calculated based on the maximum uranium content of 12.4 ppm in the garnet biotite gneiss of the region as reported by Ray et al. (2008) and assuming a secular equilibrium between the activities of ^{238}U and ^{226}Ra in the rock); *E*—0.03; *d*—2500 kg/m^3.

For the weathered lateritic aquifer (midland)—A_{Ra}—32.3 Bq/kg (calculated based on the average uranium content of 2.6 ppm in the granitic gneiss of KKB as reported by Braun (2006)); *E*—0.13; *d*—2500 kg/m^3.

For the sandy alluvial aquifer (lowland), radium content A_{Ra} was58.5 Bq/kg (the average activity in this area); emanation coefficient *E*—0.009 (Todd and Mays 2005); *d*—2500 kg/m^3.

The calculated representative values of radon production rate for the fractured granitic aquifer (high land), the weathered lateritic aquifer (midland), and the sandy alluvial aquifer (low land) were 87.89, 79.88, and 10.01 Bq m^{-3} h$^{-1,}$ respectively. The lowest radon production potential was observed in the sandy alluvial aquifer in the lowland supporting the observed low ^{222}Rn activity in this aquifer.

In the midland region, the weathered lateritic aquifer exhibits a significant emanation coefficient of 0.13, indicating substantial weathering and tectonic deformations. This high emanation coefficient is a distinctive feature of reservoir rocks with a dense crack network and/or high porosity, as well as zones that have been subjected to intense brittle tectonic deformation (Przylibski 2011). Notably, parameters such as the concentration of $^{238}U/^{226}Ra$ in rocks and emanation coefficient are less susceptible to climatic and anthropogenic influences when compared to ^{222}Rn activity in groundwater, soil, and air environments (Pereira et al. 2017).

(b) **Lineament density**

Lineament density has proven to be a valuable tool in identifying potential groundwater zones, as demonstrated in several studies (Martins et al. 2019; Benjmel et al. 2020; Tolche 2020). In addition to this application, previous research has suggested that lineament density is also a contributing factor to the abundance of radon in soil and groundwater, and has been utilized to model radon occurrence and migration (Martins et al. 2019). Notably, the mean lineament density for fractured granitic aquifers in highland areas was found to be 1.8 km/km^2, while weathered lateritic aquifers exhibited a density of 2.3 km/km^2 and sandy alluvial aquifers had a much lower density of 0.6 km/km^2. This disparity in lineament density explains the relatively higher levels of ^{222}Rn activities observed in the former two types of aquifers compared to the latter.

3.2.9.1 Understanding the ^{222}Rn Release Mechanisms in Groundwater from Hard Rock Terrains: A Conceptual Model

Drawing upon the findings of this study, a comprehensive conceptual model is developed to elucidate the underlying mechanisms of radon anomalies in groundwater of hard rock aquifers. Figure 3.11 depicts a schematic diagram illustrating the release mechanisms of ^{222}Rn into hard rock aquifers. It is hypothesized that the presence of radon in groundwater is primarily determined by the concentration and distribution of uranium and radium in the source rock, as well as their geochemical enrichment.

Furthermore, the depth of the bedrock, direction of groundwater flow, residence time, and length of the flow all play important roles in determining the concentration and distribution of radon in groundwater. Structural features such as fractures and lineaments, as well as the degree of weathering and fracturing, and the emanation coefficient of the reservoir rock, are also critical factors influencing the spatial distribution of radon in groundwater.

From a hydrogeological perspective, the findings of this study suggest that high levels of radon activity in groundwater can indicate the presence of uranium and radium-enriched zones, while relatively high radon activity in fractured aquifers may indicate the presence of active lineaments that could serve as potential sources of drinking water.

Medium radon activity in fractured aquifers could represent wells with moderate yields, suitable for both drinking and irrigation purposes, whereas low radon activity

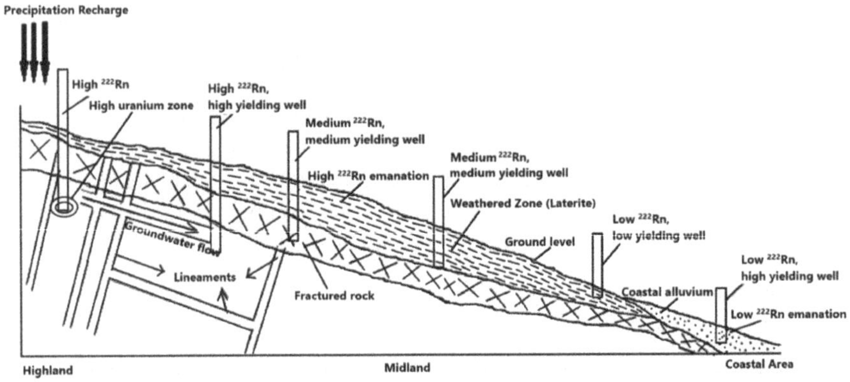

Fig. 3.11 Conceptual model illustrating the mechanism of radon release in hard rock aquifers (*Source* Sukanya et al. 2021)

in weathered, laterite, and alluvial aquifers with high depth to bedrock could represent wells with low yields, suitable for meeting the drinking water needs of individuals. Therefore, the proposed conceptual model and the hydrogeological findings of this study could serve as valuable tools for water prospecting in hard rock terrains.

3.3 Radon Distribution in Hydrothermal Systems

There exists a long history of studies based on radioactivity in hydrothermal systems (e.g., hot spring systems) in the literature (Wollenberg 1975; Horvath et al. 2000; Sakoda et al. 2008; Jalili-Majareshin et al. 2012). Most of these investigations revolve around two radioisotopes—^{226}Ra and ^{222}Rn (Girault and Perrier 2014; Tabar and Yakut 2014; Girault et al. 2018; Nugraha et al. 2021; Zemour et al. 2022). ^{222}Rn distribution in hydrothermal systems primarily depends on the bedrock geology, solubility of the parent isotope (^{226}Ra), and alpha recoil mechanisms. Also, the solubility coefficient (S) of radon in water is a function of water temperature. The empirical equation for radon solubility coefficient is (Koike et al. 2014)

$$S = 0.1057 + 0.405\exp(-0.0502 \times T) \tag{3.3}$$

This temperature effect on ^{222}Rn activity is generally observed in water with temperature differences of 30–50 °C (Tabar and Yakut 2014; Sukanya et al. 2021). Zhang et al. (2022) observed an increase in radon activity concentration (117–285 Bq/L) with respect to decreasing temperature (87.9 °C–85 °C). However, the calculated solubility coefficients suggested that this radon anomaly was not related to solubility-temperature dependency. Based on the equation, it was found that there is a very little difference in radon gas solubility (=0.7% higher) at 85 °C than that at

87.9 °C. Hence, it was inferred that the radon anomalies were related to some other factors like seismic activity.

Mixing of deep hot water with shallow cold water is another factor determining the radon activity concentrations in hot springs. In some deep sea hydrothermal systems, mixing of hot hydrothermal fluids with cold seawater drawn into subsurface has been reported (Bemis et al. 2012; Moore et al. 2021). In such cases, both low- and high-temperature hydrothermal fluids tend to be enriched in ^{222}Rn over cold seawater. This temporal variability of ^{222}Rn activity concentrations could be used as an important tool for scrutinizing sub-seafloor processes (Cochran and Kadko 2008).

3.4 Radon Distribution in River Water

Radon activity in surface water is low as radon is readily lost to the atmosphere through gas transfer. Water that enters the subsurface increases in radon activity over a period of around 20 days until equilibrium between radon production and decay is reached (Bourke et al. 2014). Radon distribution in river water is investigated either for estimating river water fluxes recharging adjacent riverine aquifer or groundwater fluxes contributing to the streamflow, i.e., surface water–groundwater interactions.

Understanding these interactions in a river system is essential for utilizing water resources effectively. Conventional techniques are mostly constrained as they cannot be used to determine groundwater seepage accurately. As ^{222}Rn in groundwater is 2–4 magnitudes higher than in surface water, the use of ^{222}Rn could serve as a suitable indicator of the exchange processes between surface water and groundwater components (Frei and Gilfedder 2021). This aspect is described in detail in Chap. 4. However, a case study is presented in the next Sect. (3.3.1) for understanding the distribution of ^{222}Rn in river water and identification of groundwater inflow into river using the data.

3.4.1 Case Study—Killiyar River Basin, India

Anurani et al. (2021) conducted a hydrological study of the Killiyar River (KR), Kerala, focusing on the use of radon as a tracer to understand the environmental flow. KR is the sixth tributary of the Karamana River, covering an area of 102 km2 with a length of 24 km. It is situated between latitudes 8°40′30″ N, 8°27′0″ N and longitudes 76°57′0″ E, 76°2′0″ E in Thiruvananthapuram district, Kerala State, India (Fig. 3.12). The river originates near Panavur (8°38′30.7″ N and 76°59′19.4″ E) at an elevation of 7.5–75 m above mean sea level (amsl), in the midland region, and flows in a dendritic to sub-dendritic pattern until it confluences with the Karamana River in the downstream region (08°27′23.4″ N and 76°57′32″ E), about 2.0 km from

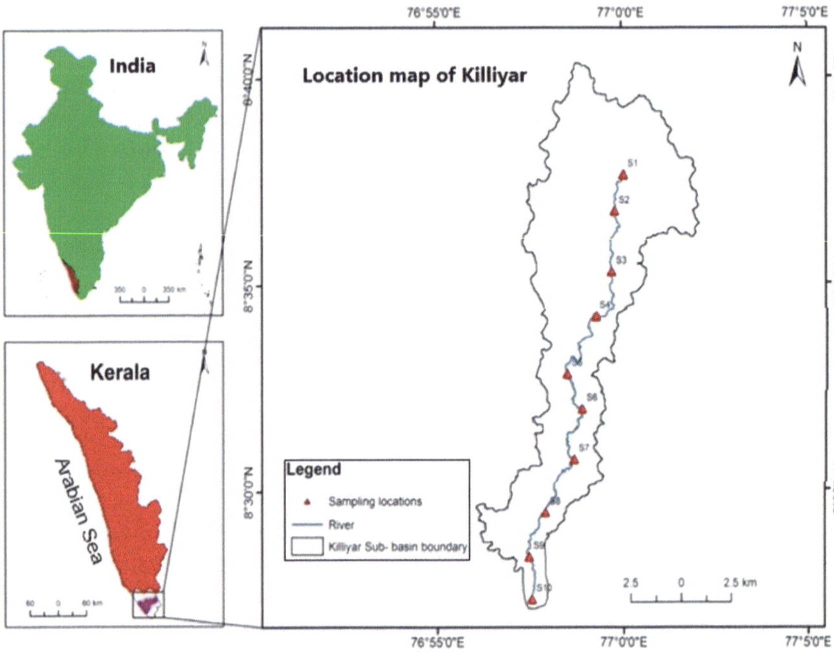

Fig. 3.12 Location map of the Killiyar River Basin, India (adapted from Anu Rani et al. 2021)

the sea coast. The geology of the river basin is dominated by garnetiferous biotite–sillimanite gneiss with or without graphite (khondalite), migmatites, and patches of charnockites, sand, and clay deposits, with 90% of the basin comprising the former.

In order to examine the environmental flow in KR, a total of 10 surface water samples were collected at intervals of approximately 3 km. The sampling was conducted on a single day during both the pre- and post-monsoon seasons in January and December 2017, respectively. The selection of the sampling sites was based on various criteria including accessibility, land use pattern, topography, and anthropogenic influences. Surface water samples were collected in 250 mL glass bottles under careful conditions to prevent atmospheric air contact. The method involved stripping the ^{222}Rn from the water by bubbling air and circulating it through a closed air loop via a desiccant tube into the ^{222}Rn counting system (RAD7, Durridge make).

The spatial distribution of ^{222}Rn activity demonstrated similar patterns during both pre- and post-monsoon seasons. The ^{222}Rn activity ranged from 157 to 4588 Bq/m^3 during pre-monsoon and from 147 to 1740 Bq/m^3 during post-monsoon. The findings revealed spatial variations of ^{222}Rn activity, with the upstream reaches of the Killiyar River (stretching from sampling sites S1–S2, within 0–3 km) exhibiting notably higher ^{222}Rn activity (>1000 Bq/m^3) compared to the downstream reaches (excluding S5) with lower ^{222}Rn activity (<500 Bq/m^3) during both seasons. However, sampling site S5, located in the mid-course of the Killiyar River (~10 km downstream of S1), also exhibited higher ^{222}Rn activity (>1000 Bq/m^3) during both seasons.

Typically, increased ^{222}Rn activities in surface water indicate the influx of radon-rich groundwater into the river. Therefore, the upstream (S1–S2; 0–3 km stretch) and midstream (S5; 10 km downstream of S1) stretches in the river, where anomalous high ^{222}Rn activities were observed, may represent zones of prospective groundwater discharge into the river. The decreasing trend of ^{222}Rn activity from upstream to downstream (excluding S5) suggests the escape of radon from surface water to the atmosphere across the boundary layer via molecular diffusion, i.e., gas exchange. Another possible explanation for this decline towards downstream is the radioactive decay of ^{222}Rn, which has a short half-life of 3.8 days.

In addition to spatial variation, temporal variation of radon activities was observed in the study area, with lower radon activity during post-monsoon season (mean = 745 Bq/m^3) compared to pre-monsoon season (mean = 1150 Bq/m^3) due to the dilution effect of monsoon and the turbulent condition of water promoting fast atmospheric escape of radon. Notably, locations with higher ^{222}Rn activities in the river course (i.e., S1–S2, 0–3 km stretch, and S5—10 km from S1) are underlain by fractured compact khondalitic rocks susceptible to weathering. These rocks lack primary porosity, and instead store groundwater in repositories with the development of secondary porosity due to weathering and fracturing, as well as other tectonic processes. The availability of groundwater in these formations is primarily determined by the rate of infiltration through fractures, which are facilitated by the foliations of these metamorphic rocks serving as planes of weakness for flow and storage.

During both seasons, the downstream reaches of the river (S9 and S10; 22–24 km stretch) exhibited the lowest ^{222}Rn activities, ranging from 157 to 312 Bq/m^3 during PRM and 147–294 Bq/m^3 during POM. This observation may be attributed to the low content of radon parent minerals in the sandy silt alluvium underlying the river bed. Additionally, the river bed in this area is composed of silty clay, which has a lower hydraulic conductivity compared to the underlying sand and silt layers.

The radon activity data allowed for the classification of groundwater potential zones within KR into three categories: high, moderate, and low or negligible potential zones, using the interpretive guide developed by Harrington et al. (2012). The high potential zones, with radon activity above 1000 Bq/m^3, are located at stretches 0–3 km upstream (S1 and S2) and 10 km from upstream (S5), indicating excellent prospects for groundwater. The moderate potential zones, with radon activity ranging between 500 and 1000 Bq/m^3, are found at the 6th km from upstream (S3) and between 14 and 16th km from upstream (S6 and S7), indicating moderate groundwater potential. Finally, the low potential zones, with radon activity ranging from 100 to 500 Bq/m^3, are located at the upper reach of 8th km from upstream (S4) and in the lower segment of the river from 20 to 24 km (S8–S10), indicating poor groundwater potential.

To explore and confirm these groundwater potential zones, geophysical (Resistivity) field surveys can be conducted. The results of these surveys can be used to confine the groundwater potential in these zones, and management options can be developed to preserve the groundwater potential for sustainable environmental flow of the river during dry season.

3.5 Conclusion and Outlook

The complex interplay between various geological factors and the spatial distribution of radon in groundwater is extensively investigated in this chapter. The parent material ^{226}Ra content, bedrock emanation coefficient, and structural features like fractures and lineaments in the parent rock all play an integral role in determining the radon distribution in groundwater. Furthermore, the residence time of groundwater and the length of radon transport along the flow path also significantly influence the radon distribution in groundwater. By utilizing the ^{222}Rn in groundwater as a proxy, it is possible to identify weak zones such as faults, lineaments, and fractures which act as significant conduits/repositories of groundwater in hard rock areas. This knowledge can be harnessed to explore radon in groundwater/soil gas as an effective tool for groundwater prospecting in hard rock terrains. This chapter emphasizes the remarkable potential of radon as a geochemical tracer for the identification and characterization of subsurface flow paths, thereby enhancing our understanding of the hydrogeological processes in hard rock areas.

In hydrological systems, radon distribution profiles in streams and rivers can serve as a valuable tool to identify groundwater inflow and river water outflow processes. This information can aid local authorities and planners in implementing integrated and holistic river management programs for sustainable environmental flow, particularly during pre-monsoon and summer seasons with community involvement.

Given its exceptional sensitivity to abrupt changes in the subsurface environment, radon holds enormous potential as a tracer in various unexplored applications. For example, radon can provide essential information on hydraulic processes such as subsurface erosion and soil piping in areas where macropore development has already occurred.

Radon can potentially be used for landslide studies, as the release of radon gas from the ground can be affected by changes in soil moisture and permeability, which can be related to landslide activity. However, the use of radon as a tool for landslide studies is still relatively unexplored and requires further research to fully understand the relationships between radon release and landslide activity. Additionally, other factors such as geology, topography, and climate can also influence radon release and complicate the interpretation of radon measurements in the context of landslides. Therefore, while radon has potential as a tool for landslide studies, it should be used in conjunction with other monitoring methods and interpreted carefully in the context of local geological and environmental conditions.

Climate research represents another promising area of interest for radon tracers. Geological carbon storage (GCS) is one of the most innovative fields of study for addressing climate change. Radon is particularly well suited for tracing the migration of gases and liquids, including greenhouse gases like CO_2 and CH_4, in porous media systems. Radon may provide valuable insights into the movement of plumes, containment, leakage rates, and trapping techniques.

Appendix

See Table 3.3.

Table 3.3 ^{222}Rn activity and physicochemical parameters of groundwater collected from the Karamana River Basin (KRB), India (*data from* Sukanya et al. 2021)

Sample ID	Non-monsoon				Monsoon			
	Temp (°C)	Depth to water table (m)	Alkalinity (mg/L)	^{222}Rn (Bq/m^3)	Temp (°C)	Depth to water table (m)	Alkalinity (mg/L)	^{222}Rn (Bq/m^3)
G-1	23.9	3	40	68,352	23	1.2	30	19,159
G-2	24.9	6	40	10,509	24.2	3.4	30	5600
G-3	25.4	5	60	5420	24.8	3.7	40	1980
G-4	25.9	5	10	1377	24.9	3	10	1080
G-5	25.9	9	40	2950	25	5.8	40	1597
G-6	25.6	8	40	3409	24.7	4.4	30	1341
G-7	26.0	9	40	464	25.4	5.4	20	268
G-8	27.5	6	80	1451	26.6	4.3	60	1357
G-9	26.5	12	60	5914	25.8	5.6	50	2986
G-10	27.5	12	60	3253	26.6	7.9	50	1320
G-11	26.6	5	60	5526	25.8	2.3	50	2189
G-12	26.7	10	10	3203	25.7	5.7	10	1180
G-13	26.9	7	40	7021	26.1	3.5	40	2870
G-14	28.1	9	60	2970	28.1	4.4	40	816
G-15	27.6	9	40	2351	26.9	5.3	30	1189
G-16	27.9	8	40	1341	27.1	5.1	30	758
G-17	28.6	6	40	970	27.9	3.3	30	577
G-18	30.6	12	40	8280	29.4	6.9	30	3458
G-19	29.0	11	40	4627	28.8	6.4	40	1867
G-20	27.8	5	60	6438	27.2	2.2	50	3523
G-21	28.8	21	40	3444	27.9	15.2	40	1189
G-22	28.2	13	40	3951	27.7	6.2	40	2389
G-23	27.4	9	40	801	27.1	3.3	30	559
G-24	27.7	16	60	3135	27.2	9.6	50	1889
G-25	21.7	4	60	1694	21.5	1.6	40	712

(continued)

Table 3.3 (continued)

Sample ID	Non-monsoon				Monsoon			
	Temp (°C)	Depth to water table (m)	Alkalinity (mg/L)	^{222}Rn (Bq/m^3)	Temp (°C)	Depth to water table (m)	Alkalinity (mg/L)	^{222}Rn (Bq/m^3)
G-26	27.6	5	40	6122	27.3	2.1	40	2587
G-27	27.4	10	20	57,527	26.8	4.4	10	34,595
G-28	28.1	10	60	8876	27.6	8.5	40	3467
G-29	26.0	7	20	167	25.6	4.2	10	92
G-30	25.7	9	40	168	25.2	4.6	30	67
G-31	28.3	6	60	3092	27.6	1.3	40	875
G-32	27.2	6	60	8110	26.8	2.4	50	3459
G-33	26.5	7	60	1114	26.2	4.2	40	668
G-34	27.8	4	80	3778	27.1	1.9	70	1120
G-35	28.0	20	40	1471	27.5	11.1	40	998
G-36	28.7	8	40	4480	28.2	5.2	30	2290
G-37	29.7	7	100	8193	29.1	4.4	90	5618
G-38	29.1	12	60	4909	28.6	5.7	50	1473
G-39	27.6	4	320	544	27.4	2.2	300	121
G-40	28.9	4	80	2714	28.4	2.3	70	1086
G-41	29.0	4	100	1204	28.6	1.8	90	817
G-42	30.9	6	60	20,793	30.1	2.7	40	9780
G-43	29.3	8	100	14,387	28.8	3.1	100	6580
G-44	28.5	7	160	9537	28.1	3.4	150	3459
G-45	28.9	12	60	881	28.2	5.8	50	211
G-46	27.9	11	40	5640	27.3	6.1	30	1471
G-47	28.0	6	60	516	27.7	3.7	40	118
G-48	26.9	4	360	3171	26.2	1.8	340	1186
G-49	27.3	10	40	1555	26.8	10.2	30	827
G-50	30.4	10	40	9370	29.8	7.2	40	3192
G-51	28.7	11	60	5384	28.2	6.3	50	1114
G-52	29.1	12	360	25,588	28.6	6.6	340	10,370
G-53	29.0	10	60	6312	28.6	3.3	60	1497
G-54	28.0	7	60	4016	27.8	2.9	40	1204
G-55	28.7	4	80	2138	28.6	1.9	70	881
G-56	27.9	3	180	317	27.2	1.9	170	149
G-57	31.1	6	320	2074	29.8	3.5	310	811

(continued)

Table 3.3 (continued)

Sample ID	Non-monsoon				Monsoon			
	Temp (°C)	Depth to water table (m)	Alkalinity (mg/L)	^{222}Rn (Bq/m^3)	Temp (°C)	Depth to water table (m)	Alkalinity (mg/L)	^{222}Rn (Bq/m^3)
G-58	28.7	6	120	2362	28.5	2.1	110	1043
G-59	28.5	7	100	2698	28.2	3.4	90	1156
G-60	28.3	8	60	10,632	27.9	4.5	60	3203
G-61	27.1	10	40	870	26.8	7.3	40	118
G-62	27.7	5	80	33,043	27.1	1.9	70	17,388
G-63	28.3	7	60	22,814	27.8	4.9	60	13,493
G-64	28.5	3	140	3081	28	1.9	120	814
G-65	28.0	4	140	473	27.8	2.1	130	115
G-66	28.3	1	180	3744	27.6	1.1	160	1182
G-67	28.0	5	40	956	27.3	2.4	30	317
G-68	31.7	25	40	6496	29.8	18.6	30	6312
G-69	28.0	11	100	1317	27.5	4.9	90	511
G-70	30.0	19	60	29,929	29.8	11.5	50	18,312
G-71	27.6	9	100	6567	27	5.4	90	1377

References

Akingboye AS, Ogunyele AC, Jimoh AT, Adaramoye OB, Adeola AO, Ajayi T (2021) Radioactivity, radiogenic heat production and environmental radiation risk of the Basement Complex rocks of Akungba-Akoko, southwestern Nigeria: insights from in situ gamma-ray spectrometry. Environ Earth Sci 80(6):1–24

Anurani GR, Lakshmi R, Joseph S (2021) Evaluation of radon (^{222}Rn) distribution and its implications vis-a-vis water quality of Killiyar River, Kerala, India. Curr World Environ 1:94–104

Appleton JD, Jones DG, Miles JCH, Scivyer C (2020) Radon gas hazard. Geol Soc Lond Eng Geol Spcc Publ 29(1):433–456

Arabi AS, Tukur Z, Funtua II, Ali MA, Kurowska E, Abdulhamid MS, Murtala AS (2021). Radon dynamics, activity concentration, and radiological risks associated with groundwater around tremor-prone Federal Capital Territory and environs, north-central Nigeria. Arab J Geosci 14(23):1–12

Belgacem A, Souid F, Telahigue F, Kharroubi A (2015) Temperature and radon-222 as tracer of groundwater flow: application to El Hamma geothermal aquifer system, southeastern Tunisia. Arab J Geosci 8(12):11161–11174

Bemis K, Lowell RP, Farough A (2012) Diffuse flow: on and around hydrothermal vents at mid-ocean ridges. Oceanography 25(1):182–191

Benjmel K, Amraoui F, Boutaleb S, Ouchchen M, Tahiri A, Touab A (2020) Mapping of groundwater potential zones in crystalline terrain using remote sensing, GIS techniques, and multicriteria data analysis (case of the Ighrem region, Western Anti-Atlas, morocco). Wat 12:471

Bhosale D, Kumar CP (2002) Simulation of seawater intrusion in Ernakulam coast. In: Proceeding of international conference on hydrology and watershed management, Hyderabad, pp 390–399

Bonczyk M, Chałupnik S, Howaniec N, Smoliński A, Wysocka M (2019) Testing device for radon migration experiments, the construction and preliminary results. Pure Appl Geophys 176(6):2557–2564

Bourai AA, Aswal S, Kandari T, Kumar S, Joshi V, Sahoo BK, Ramola RC (2016) Study of radon flux from soil in Budhakedar region using SRM. Radiat Prot Dosimetry 171(2):267–270

Bourke SA, Cook PG, Shanafield M, Dogramaci S, Clark JF (2014) Characterisation of hyporheic exchange in a losing stream using radon-222. J Hydrol 519:94–105

Braun I (2006) Pan-African granitic magmatism in the Kerala Khondalite Belt, southern India. J Asian Earth Sci 28(1):38–45

Chacko T, Kumar GR, Newton RC (1987) Metamorphic PT conditions of the Kerala (South India) khondalite belt, a granulite facies supracrustal terrain. J Geol 95(3):343–358

Cho BW (2017) Radon concentrations in groundwater of the Goesan Area, Korea. J Soil Groundwater Environ 22(5):63–70

Cho BW, Choo CO (2019) Geochemical behavior of uranium and radon in groundwater of Jurassic granite area, Icheon, Middle Korea. Water 11(6):1278

Cinelli G, De Cort M, Tollefsen T, Achatz M, Ajtić J, Ballabio C, Barnet I, Bochicchio F, Borelli P, Bossew P, Braga R (2019) Radon-Chapter 5. In: European atlas of natural radiation. Publication Office of the European Union

Cinti D, Vaselli O, Poncia PP, Brusca L, Grassa F, Procesi M, Tassi F (2019) Anomalous concentrations of arsenic, fluoride and radon in volcanic-sedimentary aquifers from central Italy: quality indexes for management of the water resource. Environ Pollut 253:525–537

Cochran JK, Kadko DC (2008) Uranium-and thorium-series radionuclides in marine groundwaters. Radioact Environ 13:345–382

Durrani SA, Ilic R (eds) (1997) Radon measurements by etched track detectors: Applications in radiation protection, earth sciences and the environment. World Sci Publ Co., Ltd., London

Fleischer RL, Mogro-Campero A (1978) Mapping of integrated radon emanation for detection of long-distance migration of gases within the earth: techniques and principles. J Geophys Res Solid Earth 83(B7):3539–3549

Frei S, Gilfedder BS (2021) Quantifying residence times of bank filtrate: a novel framework using radon as a natural tracer. Water Res 201:117376

Galiana-Merino JJ, Molina S, Kharazian A, Toader VE, Moldovan IA, Gómez I (2022) Analysis of Radon Measurements in Relation to Daily Seismic Activity Rates in the Vrancea Region, Romania. Sensors 22(11):4160

Giletti BJ, Kulp JL (1955) Radon leakage from radioactive minerals. Ame Mineral: J Earth Planet Mater 40(5–6):481–496

Gilfedder BS, Frei S, Hofmann H, Cartwright I (2015) Groundwater discharge to wetlands driven by storm and flood events: Quantification using continuous Radon-^{222} and electrical conductivity measurements and dynamic mass-balance modelling. Geochim Cosmochim Acta 165:161–177

Girault F, Perrier F, Przylibski TA (2018) Radon-^{222} and radium-226 occurrence in water: a review. Geol Soc Lond Spec Publ 451(1):131–154

Girault F, Perrier F (2014) The Syabru-Bensi hydrothermal system in central Nepal: 2. modeling and significance of the radon signature. J Geophys Res Solid Earth 119(5):4056–4089

Gruber V, Bossew P, De Cort M, Tollefsen T (2013) The European map of the geogenic radon potential. J Radiol Prot 33:51–60

Gundersen LC, Schumann RR, Otton JK, Dubiel RF, Owen DE, Dickinson KA (1992) Geology of radon in the United States. Geol Soc Am Spec Pap 271:1–16

Harrington N, Noordijn S, Cook P (2012) Evaluation of approaches to modelling surface water-groundwater interactions around drains in the South East of South Australia. Phase 1. Goyder Inst Water Res Tech Rep Ser 12(1)

Heriawan MN, Syafi'i AA, Saepuloh A, Kubo T, Koike K (2021) Detection of near-surface permeable zones based on spatial correlation between radon gas concentration and DTM-derived lineament density. Nat Resour Res 30(4):2989–3015

Horvath A, Bohus LO, Urbani F, Marx G, Piroth A, Greaves ED (2000) Radon concentrations in hot spring waters in northern Venezuela. J Environ Radioact 47(2):127–133
Idriss H, Salih I, Sam A (2011) Study of radon in ground water and physicochemical parameters in Khartoum state. J Radioanal Nucl Chem 290(2): 333–338
Ioannides K, Papachristodoulou C, Stamoulis K, Karamanis D, Pavlides S, Chatzipetros A, Karakala E (2003) Soil gas radon: a tool for exploring active fault zones. Appl Radiat Isot 59(2–3):205–213
Jacob N, Babu DS, Shivanna K (2009) Radon as an indicator of submarine groundwater discharge in coastal regions. Curr Sci 97:1313–1320
Jalili-Majareshin A, Behtash A, Rezaei-Ochbelagh D (2012) Radon concentration in hot springs of the touristic city of Sarein and methods to reduce radon in water. Radiat Phys Chem 81(7):749–757
Kalip A, Haque MF, Gaiya S (2018) Estimation of annual effective dose due to ingestion and inhalation of radon in groundwater from Kaduna, Nigeria. Phys Sci Int J 19(3):1–12
Knutsson G, Olofsson B (2002) Radon content in groundwater from drilled wells in the Stockholm region of Sweden. Norg Geol Under Bull 439:79–85
Koike K, Yoshinaga T, Asaue H (2014) Characterizing long-term radon concentration changes in a geothermal area for correlation with volcanic earthquakes and reservoir temperatures: a case study from Mt. Aso, southwestern Japan. J Volcanol Geoth Res 275: 85–102
Kumar A, Sharma S, Mehra R, Kanwar P, Mishra R, Kaur I (2018) Assessment of radon concentration and heavy metal contamination in groundwater of Udhampur district, Jammu & Kashmir, India. Environ Geochem Health 40(2):815–831
Lekshmi R, Arunima S, Jojo PJ (2018) Determination of radon exhalation rates and emanation factor of some soil samples collected from Southern Seashore of Kerala, India. J Ultra Sci Phys Sci (JUSPS-A) 30:80
Manisa K, Erdogan M, Zedef V, Bircan H, Biçer A (2022) Variations of 222Rn concentrations over active fault system in Simav, Kütahya, Western Turkey: Possible causes for soil-gas ^{222}Rn anomalies. Appl Radiat Isot 190:110484
Martins L, Pereira A, Oliveira A, Fernandes A, Sanches Fernandes LF, Pacheco FAL (2019) An assessment of groundwater contamination risk with radon based on clustering and structural models. Water 11(5):1107
Mercadier J, Annesley IR, McKechnie CL, Bogdan TS, Creighton S (2013) Magmatic and metamorphic uraninite mineralization in the western margin of the Trans-Hudson orogen (Saskatchewan, Canada): a uranium source for unconformity-related uranium deposits? Econ Geol 108(5):1037–1065
Moore WS, Frankle JD, Benitez-Nelson CR, Früh-Green GL, Lang SQ (2021) Activities of 223Ra and ^{226}Ra in fluids from the Lost City hydrothermal field require short fluid residence times. J Geophys Res Oceans 126(12):e2021JC017886
Nagabhushana SR, Suresh S, Sannappa J, Srinivasa E (2020) Study on activity of radium, radon and physicochemical parameters in ground water and their health hazards around Tumkur industrial area. J Radioanal Nucl Chem 323(3):1393–1403
Nandakumaran P, Vinayachandran N (2020) A preliminary appraisal of radon concentration in groundwater from the high background radiation area (HBRA) of Coastal Kerala. J Geol Soc India 95(5):491–496
Naskar AK, Gazi M, Barman C, Chowdhury S, Mondal M, Ghosh D, Sinha B, Deb A (2018) Estimation of underground water radon danger in Bakreswar and Tantloi Geothermal Region, India. J Radioanal Nucl Chem 315(2):273–283
Nguyen PTH, Vu NB, Le Cong H (2018) Soil radon gas in some soil types in the rainy season in Ho Chi Minh City, Vietnam. J Environ Radioact 193:27–35
Niranjan RS, Ningappa C, Yashaswini T, Chamaraja NA, Rangaswamy DR, Sannappa J (2017) Studies on radon concentration in drinking water around Hemavathi river basin, Karnataka State, India. J Radioanal Nucl Chem 314(1):321–331

Nugraha ED, Hosoda M, Mellawati J, Untara U, Rosianna I, Tamakuma Y, Modibo OB, Kranrod C, Kusdiana K, Tokonami S (2021) Radon activity concentrations in natural hot spring water: dose assessment and health perspective. Int J Environ Res Public Health 18(3):920

Ogden AE, Welling WB, Funderburg RD, Boschult LC (2020) A preliminary assessment of factors affecting radon levels in Idaho. In: Radon, radium, and other radioactivity in ground water. CRC Press, pp 83–96

Pereira A, Lamas R, Miranda M, Domingos F, Neves L, Ferreira N, Costa L (2017) Estimation of the radon production rate in granite rocks and evaluation of the implications for geogenic radon potential maps: a case study in Central Portugal. J Environ Radioact 166:270–277

Pinto PV, Kumara KS, Karunakara N (2020) Mass exhalation rates, emanation coefficients and enrichment pattern of radon, thoron in various grain size fractions of monazite rich beach placers. Radiat Meas 130:106220

Pommé S (2015) Typical uncertainties in alpha-particle spectrometry. Metrologia 52(3):S146

Przylibski TA (2011) Shallow circulation groundwater–the main type of water containing hazardous radon concentration. Nat Hazard 11(6):1695–1703

Przylibski TA (2015) Radon research in Poland: a review. Solid State Phenom 238:90–115

Przylibski TA (2018a) Radon. A radioactive therapeutic element. In: Gillmore GK, Perrier FE, Crockett RGM (eds) Radon, health and natural hazards, vol 451. Geological Society, London, Special Publications, pp 209–236

Przylibski TA (2018b) Radon: a radioactive therapeutic element. Geol Soc Lond Spec Publ 451(1):209–236

Przylibski TA, Gorecka J (2014) ^{222}Rn activity concentration differences in groundwaters of three Variscan granitoid massifs in the Sudetes (NE Bohemian Massif, SW Poland). J Environ Radioact 134:43–53

Przylibski TA, Domin E, Gorecka J, Kowalska A (2020) ^{222}Rn concentration in groundwaters circulating in granitoid massifs of Poland. Water 12(3):748

Przylibski TA, Staśko S, Domin E (2022) Radon groundwater in a radon-prone area: possible uses and problems: an example from SW part of Kłodzko Valley, Sudetes, SW Poland. Environ Geochem Health 1–17

Przylibski TA (2005) Radon, specific component of medicinal waters in the Sudety Mountains. Oficyna Wydawnicza Politechniki Wrocławskiej, Wrocław, (in Polish)

Raju GK, Mathai J, Kumar GR, Nair NGK (1986) Natural radioactivity distribution studies in Trivandrum district, Kerala, India. Proc Indian Acad Sci-Earth Planet Sci 95(3):397–407

Ramsiya M, Joseph A, Eappen KP, Visnuprasad AK (2019) Activity concentrations of radionuclides in soil samples along the coastal areas of Kerala, India and the assessment of radiation hazard indices. J Radioanal Nucl Chem 320(2):291–298

Ravikumar P, Davis D, Mathew S, Somashekar RK, Prakash KL (2014) Spatio-temporal variation in radon concentration in groundwater with respect to rock types: a case study from Chitradurga district, Karnataka. J Geol Soc India 83(2):156–164

Ray L, Roy S, Srinivasan R (2008) High radiogenic heat production in the Kerala Khondalite block, Southern Granulite province, India. Int J Earth Sci 97(2):257–267

Robertson A, Allen J, Laney R, Curnow A (2013) The cellular and molecular carcinogenic effects of radon exposure: a review. Int J Mol Sci 14(7):14024–14063

Sakoda A, Ishimori Y (2014) Calculation of temperature dependence of radon emanation due to alpha recoil. J Radioanal Nucl Chem 299(3):2013–2017

Sakoda A, Hanamoto K, Ishimori Y, Nagamatsu T, Yamaoka K (2008) Effects of some physical conditions on leaching rate of radon from radioactive minerals originating from some hot springs. Radiat Meas 43(1):106–110

Singaraja C, Chidambaram S, Jacob N, Selvam S, Johnsonbabu G, Anandhan P (2016) Radon levels in groundwater in the Tuticorin district of Tamil Nadu, South India. J Radioanal Nucl Chem 307:1165–1173

Schaper JL, Zarfl C, Meinikmann K, Banks EW, Baron S, Cirpka OA, Lewandowski J (2022) Spatial variability of radon production rates in an alluvial aquifer affects travel time estimates of groundwater originating from a losing stream. Water Resour Res 58(4):e2021WR030635

Scheib C, Appleton JD, Miles JCH, Hodgkinson E (2013a) Geological controls on radon potential in England. Proc Geol Assoc 124:910–928

Scheib C, Appleton JD, Miles JCH, Hodgkinson E (2013b) Geological controls on radon potential in England. Proc Geol Assoc 124(6):910–928

Sharma DA, Keesari T, Rishi MS, Thakur N, Pant D, Sangwan P, Sahoo BK, Kishore N (2020) Distribution and correlation of radon and uranium and associated hydrogeochemical processes in alluvial aquifers of northwest India. Environ Sci Pollut Res 27(31):38901–38915

Soman K (1987) Geology of Kerala. Geological Survey of India, Bangalore

Soman K (2002) Geology of Kerala. J Geol Soc India, Bangalore 335

Somashekar R, Ravikumar P (2010) Radon concentration in groundwater of Varahi and Markandeya river basins, Karnataka State, India. J Radioanal Nucl Chem 285(2):343–351

Sukanya S, Noble J, Joseph S (2021) Factors controlling the distribution of radon (^{222}Rn) in groundwater of a tropical mountainous river basin in southwest India. Chemosphere 263:128096

Tabar E, Yakut H (2014) Radon measurements in water samples from the thermal springs of Yalova basin, Turkey. J Radioanal Nucl Chem 299(1):311–319

Tallini M, Parisse B, Petitta M, Spizzico M (2013) Long-term spatio-temporal hydrochemical and ^{222}Rn tracing to investigate groundwater flow and water–rock interaction in the Gran Sasso (central Italy) carbonate aquifer. Hydrogeol J 21(7):1447–1467

Tanner AB (1964) Radon migration in the ground: a review. Nat Radiat Environ 161–190

Terray L, Gauthier PJ, Breton V, Giammanco S, Sigmarsson O, Salerno G, Caltabiano T, Falvard A (2020) Radon activity in volcanic gases of Mt. Etna by passive dosimetry. J Geophys Res Solid Earth 125(9):p.e2019JB019149

Thivya C, Chidambaram S, Thilagavathi R, Keesari T, Rao MS, Prasanna MV, Nepolian M (2015) Occurrence of high uranium and radon in hard rock aquifers of South India-evaluating the temporal and spatial trends. Groundw Sustain Dev 1(1–2):68–77

Todd DK, Mays LW (2005) Groundwater hydrology, 3rd edn. Wiley, New York

Tolche AD (2020) Groundwater potential mapping using geospatial techniques: a case study of Dhungeta-Ramis sub-basin, Ethiopia. Geol Ecol Landsc 5:1–16

United Nations Scientific Committee on the Effects of Atomic Radiation (2009) UNSCEAR 2006 report, vol II, annex E: sources-to-effects assessment for radon in homes and workplaces. United Nations, New York

United Nations Scientific Committee on the Effects of Atomic Radiation (2000) UNSCEAR 2000 report, vol I: sources. United Nations

Watson RJ, Smethurst MA, Ganerød GV, Finne I, Rudjord AL (2017) The use of mapped geology as a predictor of radon potential in Norway. J Environ Radioact 166:341–354

WHO (2009) WHO handbook on indoor radon: a public health perspective. World Health Organization, France

Wollenberg HA (1975) Radioactivity of geothermal systems (No. LBL-3232; CONF-750525-12). California Univ., Berkeley (USA). Lawrence Berkeley Lab

Wood WW, Kraemer TF, Shapiro A (2004) Radon (^{222}Rn) in ground water of fractured rocks: a diffusion/ion exchange model. Groundwater 42(4):552–567

Ye YJ, Xia XQ, Dai XT, Huang CH, Guo Q (2019) Effects of temperature, salinity, and pH on ^{222}Rn solubility in water. J Radioanal Nucl Chem 320(2):369–375

Zemour Y, Mebrouk N, Mayer A, Mekebret I, Sherif MI (2023) Hydrochemical and geological controls on dissolved radium and radon in northwestern Algeria hydrothermal groundwaters. Chemosphere 313:137573

Zhang S, Shi Z, Wang G, Yan R, Zhang Z (2022) Application of the extreme gradient boosting method to quantitatively analyze the mechanism of radon anomalous change in Banglazhang hot spring before the Lijiang Mw 7.0 earthquake. J Hydrol 612:128249

Chapter 4
Radon in Surface Water–Groundwater Interaction Studies

4.1 Introduction

For several centuries, groundwater and surface water were considered as separate entities in terms of water resource management. But this concept turned invalid when later studies from the last century proved that these two inextricable components interact in a variety of ways depending on physiographic settings (Winter et al. 1998; Foster and Allen 2015). The probable interconnection conditions within a river–aquifer system are (a) either connected, (b) transitional, or (c) disconnected (Rivière et al. 2014; Koehn et al. 2019; Fuchs et al. 2019; Xian et al. 2021).

The interconnection between groundwater and surface water bodies remains poorly understood in many river basins throughout the world, and yet is fundamental for effective water resource management. For pragmatic application of groundwater–river interactions on water resource sustainability, knowledge on factors influencing groundwater and river water interactions is necessary. This chapter provides an overview of river water–groundwater interactions in KRB.

4.2 Physical Interaction

The physical interaction of surface water and groundwater gained a phenomenal attention when Meinzer (1923) put forth his findings on the different ways by which rivers and aquifers interact. There are mainly three ways in which these two components interact:

(1) **Gaining Streams**: Gaining or effluent streams are those receiving water from a groundwater system (Fig. 4.1) through the zone of saturation (Sahoo and Sahoo 2019). This occurs at locations where the stage of river/stream is lower than the piezometric level of the aquifer (Stefania et al. 2018).

(2) **Losing Streams**: In a losing or influent stream, groundwater in the aquifer system adjacent to the river/stream is recharged by river water (Hintze et al.

S. S. and S. Joseph, *Environmental Radon*, Environmental Science and Engineering,
https://doi.org/10.1007/978-981-99-2672-5_4

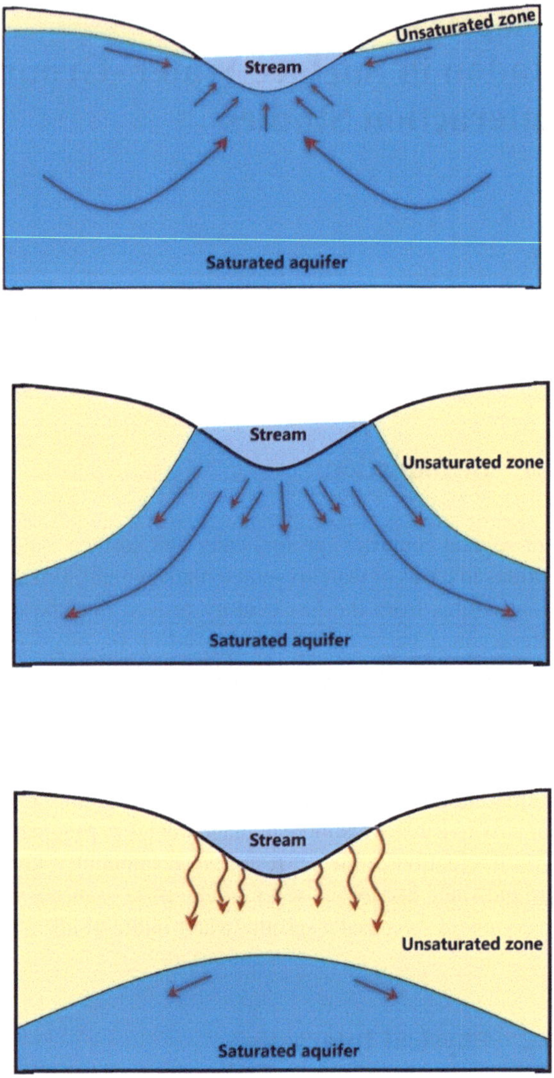

Fig. 4.1 Conceptual diagrams of gaining stream, losing stream, and disconnected stream (after Winter et al. 1998)

2020). This occurs at locations where stream stage is higher than the water table, i.e., piezometric level of aquifer (Fig. 4.1).

(3) **Combination of gaining and losing streams**: Rivers gaining water in some reaches and losing water in other reaches, or gaining and losing in the same river reach at different times.

In cases where the influent stream is separated from the underlying groundwater by means of zone of aeration, it is termed a "perched stream". More recently, the term "perched stream" is often interchangeably used as "disconnected stream", whereas the term "connected stream" is used for influent streams losing water to adjacent groundwater system through saturation zone, i.e., continuous saturate zone (Fig. 4.1). Water movement from the river into aquifer is restricted mainly due to poor hydraulic connection. Quichimbo et al. (2020) observed that infiltration rates within a disconnected system are higher than those coming under a connected flow regime. If the streambed-induced recharge rate is higher than the lateral groundwater recharge, the water table in the disconnected stream may have a visible mound below the stream (Winter et al. 1998).

Winter et al. (1998) classified another type of interaction termed "bank storage" where interaction occurs when water moves from river to the stream bank due to sudden rapid rise in river stage resulting from intense precipitation events, release of water from dam, rapid snowmelt, etc. The bank storage entirely depends on stage fluctuation which is slightly different than gaining and losing streams.

This water stored in the stream bank returns to the river within a short span of few days to weeks of time. However, when the stream stage rises above the banks and spreads to large areas of the land surface, extensive recharge occurs throughout this flooded region. That being the case, the return flow of recharged floodwater to the stream may delay for months to years due to the longer groundwater flow paths.

The size of a storage zone will be significantly decreased by a less permeable streambeds, aquitards, or anisotropic aquifers having very low vertical hydraulic conductivity (Chen and Chen 2003). Depending on the frequency, magnitude, and intensity of storms and on the related magnitude of increases in stream stage, some streams and adjacent shallow aquifers will be in a continuous readjustment from interactions related to bank storage and overbank flooding (Winter et al. 1998).

4.3 SGI and Hydrochemical Dynamics

Even though groundwater and river water components have distinct chemical properties, hydrochemical evaluation cannot be dealt separately in systems influenced by river water and groundwater interactions (Biehler et al. 2020; Zhu et al. 2020). Chemically, groundwater carries higher concentrations of dissolved elements compared to river water due to its higher rates of water–rock interactions in contrast with river water. Also, river water is more prone to pollution and quickly disperses the pollutant through self-purification dilution, chemical, and biological processes. Hyporheic zone, a porous zone connecting stream water and groundwater is characterized by steep physicochemical gradients (Palumbo-Roe et al. 2012). These chemical and biological processes occurring at the hyporheic zone significantly control the hydrochemistry of groundwater and river water.

A temporal shift of biological activity and concentration gradients of dissolved oxygen, nitrate, ammonium, etc., in the subsurface is observed in hyporheic zone due to varying local surface water–groundwater interactions (Tonina 2012; Bertrand et al. 2014). Interaction-based chemical alteration of surface water and groundwater has been reported worldwide in different ecosystems. Yuan et al. (2020) identified that water transfer from the Yellow River to the Fen River in China has significantly changed the surface water–groundwater interactions resulting in water degradation in the water-receiving river.

Obvious chemical transformation of groundwater was observed due to the recharge from a polluted river in a mountain–oasis ecosystem in Central Asia located in the Central Tianshan Mountains in Xinjiang, China (Sun et al. 2022). Boester and Rüde (2020) explored the potential of gadolinium for tracing surface water–groundwater interaction in Karst ecosystems. Li et al. (2021) noted significant amounts of dissolved constituents in groundwater contributed by the vertical inflow of surface water in the Jinci Karstic system, China. Recent studies have shown that groundwater–river water exchange could play a vital role in shaping microbial communities and biogeochemical cycling within intermittent creeks and associated alluvial aquifers (Korbel et al. 2022).

In this context, determination of chemical interaction is inevitable for the realization of pollution studies in both river water and connected groundwater components.

4.4 Factors Controlling River–Groundwater Interaction

Certain factors are responsible for large-scale exchange between groundwater and river water components within a system. They include: (i) Magnitude and distribution of hydraulic conductivity within the channel and associated fluvial plain sediments, (ii) stream stage–groundwater level relationship, (iii) geometry and position of stream channel within the alluvial system, and (iv) stream bed topography and porosity (Martinez et al. 2015; Naganna et al. 2017; Balbarini et al. 2017; Robinson et al. 2022).

The direction of the hydrological exchange processes largely depends on hydraulic head, which is again controlled by rainfall events, patterns of seasons, etc. (Wang et al. 2014). However, hydraulic conductivity of soils and rocks within the system determines the water flow rate. Anthropogenic (e.g., land use, groundwater abstraction, etc.) and natural (catchment features like geology, geomorphology, etc.) synergistically impose alterations on river–groundwater hydrological exchange (Safeeq and Fares 2016; Boester and Rüde 2020).

The gradients between river and groundwater systems are generally small in geological formations having high hydraulic conductivity, whereas, in those with low hydraulic conductivity, it is vice versa. In regions bearing pronounced relief, geology of bedrock, topography, etc., could have significant control. From a geomorphological perspective, topographic slope, channel network development, drainage density, basin geometry and size, etc., are responsible for hydrological exchange.

Relocation of sediment grains on streambed may traps stream water in the sediment interstices and releases interstitial water to the stream. Additionally, pressure variations due to geomorphological features (e.g., pool riffle sequences, discontinuities, etc.) on the streambed should be taken into account while looking for small-scale exchange of surface water and groundwater components (Trauth et al. 2014; Rau et al. 2017).

4.5 Scales of Interactions

As mentioned in the previous Sect. 4.4, surface water–groundwater exchange takes place in different scales: (i) Large-scale interactions in which the whole river basin influences the interaction and (ii) small-scale or local-scale interactions within hyporheic zone chiefly controlled by stream bed features.

(i) **Catchment-scale Interactions**

At catchment scale (large/watershed scale), recharge–discharge dynamics mainly depend on the location of the river with respect to adjacent groundwater systems. Climate, catchment and aquifer internal geometry, and land use influence basin-scale interactions from an ecological point of view (Joo et al. 2018). In other words, the location of recharge or discharge zone and the quantity of water exchanged are important at this scale. According to Bertrand et al. (2014), seasonality influences the structure of biocenosis with a long metabolism cycle in a biotope.

Groundwater flow predominantly governs the hydrological exchange at regional and intermediate scales (Flipo et al. 2014). In terms of surface runoff, structural connectivity results from microtopography, and functional connectivity resulting from the spatial variability of saturated areas, infiltration properties, vegetation, and flow resistance within the system (Appels et al. 2011). The general assumption is that when the elevation of water table is higher than surface water bed, groundwater gets discharged into surface water body. However, under certain conditions, it is observed that deeper portions of river water systems are actually losing water antithetical to the aforementioned assumption (Wang et al. 2022). This type of interaction depends on the continuity of local groundwater flow system boundary. In conditions where the boundary is not continuous below river water body, river water is likely to get recharged in the adjacent aquifer in deeper portions (Lerner 2020).

The seepage to the groundwater from a surface water body tends to increase or begin with a decrease in the water-table elevation, an increase in anisotropy, hydraulic conductivity, and bathymetry; and the presence of high-permeability zones at shallow depths on the down-gradient side of surface bodies (Xiao et al. 2017; Janos et al. 2018). Configuration or curvature of water table (i.e., concavity and convexity) is another factor influencing surface water–groundwater interaction (Quichimbo et al. 2020). For instance, a considerable upward groundwater flow occurs in regions of water-table concavity. In regions where water table is convex, groundwater flux tends to have a downward flow, which again depends on topography (Haaf et al. 2020). A water table inclined or sloped towards a river water body is steeper, resulting in concave condition and groundwater movement or seepage into surface water.

(ii) **Channel Scale (Hyporheic Zone) Interactions**

Apart from large-scale interactions, surface water components interact with the water in underlying streambed sediments (Janos et al. 2018) which is independent of catchment-scale interactions. This type of interaction creates localized gaining or losing zones within a stream reach where water exchange occurs through the sediments beneath. Channel sinuosity is one of the elements influencing flux exchange between stream and sediments. Such interaction of surface water and groundwater forms a zone of intense biogeochemical activity known as hyporheic zone (Trauth et al. 2014). Characterized by biochemical reactions involving the addition/removal of nutrients and microbial abundance, hyporheic zone interactions are important for investigating studies related to water quality and ecohydrology (Trauth and Fleckenstein 2017).

Numerous physical drivers, viz. regional groundwater flow patterns (Trauth and Fleckenstein 2017; Wu et al. 2020), streambed topography (Mojarrad et al. 2019, 2021), hydraulic conductivity (Laube et al. 2022), bedform geometry (Jiang et al. 2020; Jin et al. 2022), stream channel gradient (Earon et al. 2020; Singh et al. 2022), sediment composition (Pescimoro et al. 2019), bedform sequences like pool riffle, vertical step-pool sequences (Magliozzi et al. 2018), lateral meandering (Huang and Chui 2022), etc., control the extent and degree of hyporheic exchange flux.

A more precise multi-scale classification was developed later by Hancock et al. (2005) as sediment scale (<1 m), reach scale (1–1000 m), and catchment scale (>1000 m). The sediment scale is dominated by hyporheic processes, while the reach and catchment scales are dominated by local and regional groundwater flow systems, respectively (Janos et al. 2018).

4.6 Significance of Surface Water–Groundwater Interaction (SGI) Studies

Understanding the hydrological exchange between surface water and groundwater components is important for improving water resource management as well as governance in terms of quantity and quality (Quichimbo et al. 2020). Estimation of recharge/discharge rates along a surface water body and adjacent aquifer is important from a quantitative aspect. Surface water groundwater interaction (SGI) studies can be useful in developing water quality restoration strategies and programs.

The mixing of surface water and groundwater entities imparts their physicochemical characteristics upon each other. Based on the connectivity extent, the water quality of surface- and groundwater components are interdependent (Trauth and Fleckenstein 2017). Additionally, SGI plays a significant role in riparian ecosystem functioning. Hence, investigations on surface water-groundwater flux exchanges are carried out for studying the functions related to riparian zones and groundwater-dependent ecosystems (Bertrand et al. 2014). In the wake of a more holistic approach to solve environmental issues, surface water–ground water interactions have been receiving wide attention from ecologists, policymakers, and watershed managers.

4.7 SGI Estimation Methods

Various techniques are available for the identification and estimation of surface water–groundwater interaction fluxes in a hydrogeological system. Numerical modeling is a powerful tool for assessing surface water–groundwater interactions, as it can incorporate complex hydrological processes and provide detailed predictions (Abdelhalim et al. 2019). However, the accuracy of the results depends on the availability and quality of input data, and the complexity of the model can lead to high computational demands and longer processing times (Holmes et al. 2020).

Baseflow separation is a clear-cut method for the computation of base flow at a catchment scale using numerical models or computer programs (Ding et al. 2022). However, this method is not convenient in the case of losing or highly regulated systems. Considering the baseflow component (regional groundwater inflow) alone may end up in overestimation or underestimation in systems with bank return flow or interflow (Yu et al. 2013; Saedi et al. 2022).

Geophysical methods, such as electrical resistivity tomography and ground-penetrating radar, can provide detailed information on subsurface structures and hydraulic properties useful in surface water–groundwater interaction studies (Binley et al. 2015). However, these methods are often limited by their inability to distinguish between water sources, such as surface water and groundwater, and quantify their respective contributions to a mixed hydrological system. Despite the progress made in geophysical instrumentation and modeling techniques, the interpretation of geophysical data can still be challenging, and particular attention must be paid to the applicability of geophysical tools in different geological or structural settings (McLachlan et al. 2017). In the case of geophysical techniques like ground penetrating radar, its utility in surface water–groundwater interaction studies is limited in areas with clays in the overburden. This is due to the attenuation of the radar signal, as noted in a study by Cassidy et al. (2014).

Geochemical techniques, including analysis of major ion concentrations, stable isotopes, and radiogenic isotopes (such as ^{222}Rn and ^{226}Ra), can be utilized to quantify groundwater inflows in gaining streams (Cook 2013). The rationale behind utilizing geochemical tracers for quantifying groundwater inflows is that the tracer concentration in groundwater is significantly distinct from that in river water and any heterogeneity is well characterized (Gleeson et al. 2013). However, in the case of major ion chemistry, their concentrations can be affected by mixing processes and reactions within the subsurface, making interpretation of the data difficult.

While stable isotopes (^{18}O, ^{2}H) are commonly used to estimate surface water groundwater interaction, they do have some limitations. For instance, in a study conducted by Dunn et al. (2008) on a Scottish upland catchment, the use of stable isotopes was found to have limitations in cases where the output signal is well mixed and lacks information content. This is due to the homogeneous isotope data that is produced, indicating the need for alternative tracers like radon typically having high-resolution ability to define longer residence times to effectively characterize groundwater discharge and flow processes (Leibundgut et al. 2009; Birkel and Soulsby 2015). This emphasizes the importance of choosing the appropriate tracer for surface water–groundwater interaction studies, with radon emerging as a superior option due to its unique properties (see Sect. 4.8) and advantages over stable isotopes and other techniques outlined in this section.

4.8 Radon (^{222}Rn) as SGI Tracer

^{222}Rn, with a short half-life (t1/2 = 3.83 days), is a widely utilized conservative radioactive tracer in the assessment of surface water–groundwater interactions (SGI) (Schubert et al. 2020). In groundwater systems, the presence of ^{222}Rn can be attributed to alpha recoil resulting from the decay of ^{226}Ra within the aquifer and production from ^{226}Ra in solution and/or adsorbed to solid surfaces. However, the latter source is often considered negligible and omitted from studies (Carvalho et al. 2014).

The dominant removal mechanism of ^{222}Rn in groundwater is decay. In surface water systems without recharge from groundwater, the sources of ^{222}Rn include production from ^{226}Ra in the water and sediment diffusion, both of which are typically minor (Gleeson et al. 2013). The removal mechanisms (sinks) of ^{222}Rn in surface water include decay and atmospheric loss. Accurate interpretation of results and a deeper understanding of SGI processes require consideration of these sources and removal mechanisms of ^{222}Rn.

In their recent work, Liao et al. (2021) evaluated the effectiveness of using ^{222}Rn as a tracer for tracking dynamic hyporheic exchange between groundwater and surface water during a single flood event. Their simulation and field observations demonstrated that ^{222}Rn is a superior tracer compared to conventional methods like water temperature and total dissolved solids in tracing dynamic hyporheic exchange. The study analyzed the effects of factors such as hydraulic conductivity, dispersivity, and bank topography on ^{222}Rn signals, suggesting it as a suitable tracer for groundwater and surface water interactions in dynamic river settings.

Radon (^{222}Rn) is a powerful tracer for examining interactions both from qualitative and quantitative perspectives (Cook et al. 2018). The high contrast between groundwater and surface water activities of ^{222}Rn makes it a convenient tracer of groundwater inflows into rivers, especially where the difference in major ion concentrations between groundwater and surface water is low as in many upper catchment streams.

The change in ^{222}Rn activities in a gaining stream (dCr/dx) is governed by groundwater inflow, in-stream evaporation, hyporheic exchange, degassing, and radioactive decay:

$$Q\frac{dCr}{dx} = I(Ci - Cr) + \omega ECr + Fh - kd\omega Cr - \lambda d\omega Cr \tag{4.1}$$

(Yu et al. 2013).

The equation, designated as Eq. (4.1), takes into account various factors viz., Q denoted by stream discharge ($m^3\ day^{-1}$), the activity of ^{222}Rn within the stream (Cr) in Bqm^{-3}, the distance along the flow direction (x) in m, the rate of groundwater inflow per unit stream length (I) in $m^3\ m^{-1}\ day^{-1}$, the activity of ^{222}Rn in the influent groundwater (Ci) in Bqm^{-3}, the flux of ^{222}Rn from the hyporheic zone (Fh) in $Bqm^{-1}\ day^{-1}$, the width of river surface (ω) in m, the mean stream depth (d) in m, the rate of evaporation (E) in m day^{-1}, the gas transfer coefficient (k) per day (day^{-1}), and the radioactive decay constant (λ), i.e., 0.181 day^{-1}. The groundwater inflow can be calculated by rearranging Eq. (4.1). Additionally, it can be applied to other tracers. If hyporheic and parafluvial exchange are not considered in a study, groundwater discharge can be overestimated (Cartwright and Gilfedder 2015; Cook et al. 2006).

The one-dimensional equation proposed by Yu et al. (2013) is generally applicable to most river systems. Nonetheless, a modified radon mass balance equation that takes into account the influence of tributaries was introduced by Frei et al. (2019).

$$Q\frac{dCr}{dx} = I.(Ci - Cr) + a1 - (a2 \times Ci) + \frac{Q_{trib}}{\omega_{trib}} \times (C_{trib} - Cr) - (Cr.k.\omega.h) - (Cr.\lambda.h.\omega) \tag{4.2}$$

where Q implies stream discharge ($m^3\ d^{-1}$); I is groundwater discharge into the stream ($m^3\ m^{-1}\ d^{-1}$), a_1 ($Bq\ m^{-1}\ d^{-1}$), and a_2 ($m^2\ d^{-1}$) are ^{222}Rn enrichment during hyporheic exchange; Q_{trib} represents discharge from a tributary ($m^3\ d^{-1}$); ω_{trib} denotes tributary width of at confluence point with main trunk/stem of stream (m); C_{trib} represents ^{222}Rn activity in tributary (Bq/m^3); ω is the width of stream (m); h denotes the stream depth (m); the gas transfer coefficient (k) per day (day^{-1}); and the radioactive decay constant (λ), i.e., 0.181 day^{-1}; the activity of ^{222}Rn within the stream (Cr) in Bq/m^3.

In order to calculate a_1 and a_2, the following equation is used:

$$a_1 \text{ and } a_2 = \frac{\gamma \times \theta \times h_{hyporheic} \times \omega}{\lambda \times T_m + 1} \tag{4.3}$$

where $h_{hyporheic}$ represents hyporheic zone thickness (m); T_m implies mean residence time of river water in hyporheic zone; and θ denotes sediment porosity which is dimensionless.

Adyasari et al. (2023) have recently published an in-depth review of the applications of radon as a groundwater tracer in surface waters. Specifically, they addressed the challenge of atmospheric evasion, identification of groundwater endmembers, offshore mixing loss, steady-state assumptions, and upscaling of groundwater discharge from ^{222}Rn measurements. Furthermore, the authors provided practical guidelines and open-source software, including R codes and FINIFLUX, to aid future researchers in implementing the ^{222}Rn method in their investigations.

4.9 Case Study—Karamana River Basin, India

For studying the SGI in the Karamana River Basin (KRB), about 15 profile sampling locations (P1–P15) were identified in the Karamana River from upstream to downstream at an interval of ~4.0 km. Then, from each profile location, one river water sample was collected. Further, based on each profile location, a linear series of groundwater samples (2–3 samples) were collected with a sampling spacing of ~50 m (based on availability and accessibility of dug wells) and set along a profile line on either side of the river and perpendicular to the river course in variable distances (up to say 200 m landward from the river bank).

Thus, based on the above, groundwater profile samples were collected from the near river zone (0–50 m from the river), intermediate river zone (50–100 m from the river) as well as distant river zone (100–150 m from the river) from left and right banks of the river (Fig. 5.1). Samples for two seasons of 2018, viz. non-monsoon (NON, April) and monsoon (MON, July) were collected for the SGI study. Thus, a total of 15 river water (RP1–RP15) and 56 river zone or profile groundwater samples were collected (Fig. 5.1).

For ^{222}Rn measurements, river water samples were collected in a 1-L bottle. Groundwater samples were collected from the bottom of dug wells and the samples were carefully collected in 250-mL glass bottles without coming into contact with atmospheric air. ^{222}Rn activities in water samples were measured using RAD7.H_2O portable instrument (Fig. 4.2).

4.9.1 Radon Activities in River Water and Profile (River Zone) Groundwater

The present study uses ^{222}Rn activities to determine the surface water–groundwater interaction (SGI) relationships and the contribution of base flow (groundwater) in the Karamana River (KR) from upstream to downstream.

(i) **Non-monsoon**

Along the course of the Karamana River, the ^{222}Rn activity of river water at the uppermost profile location P1 (RP1, 0 km) in highland was significantly higher (4060 Bq/m^3) during non-monsoon. This implies the groundwater contribution to the river during non-monsoon (low flow conditions). A remarkable decline in ^{222}Rn activity was found at Profile P2 (RP2, 5 km from P1) in highland compared to P1 (Table 4.1 and Fig. 4.2). Despite this fall, a significant activity (592 Bq/m^3) was noticed at this stretch. Further, there was a slightly increased activity (694 Bq/m^3) at

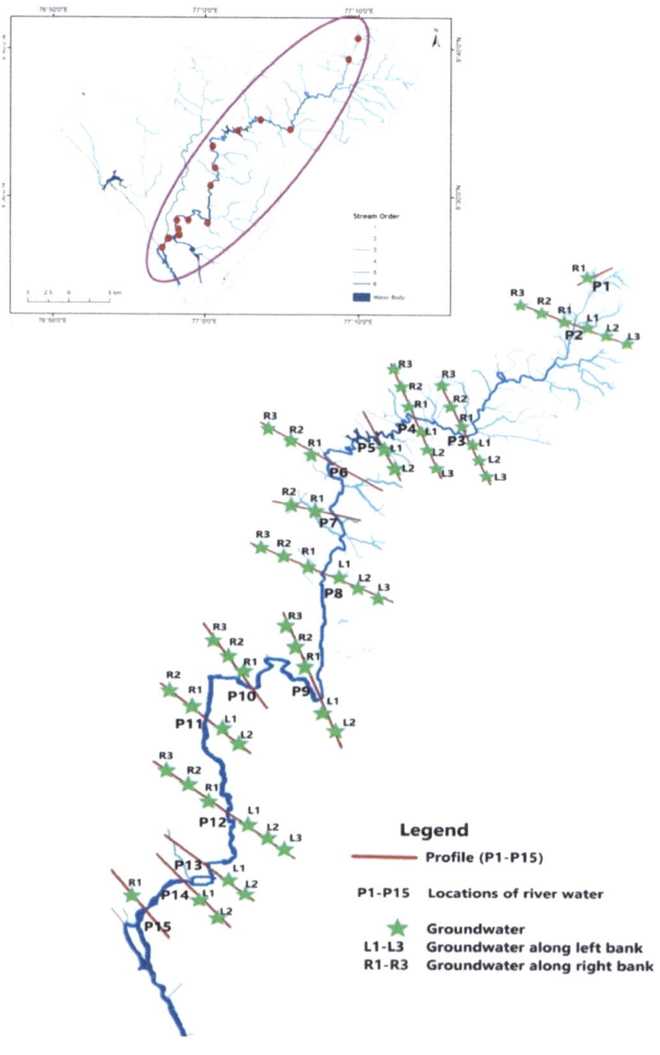

Fig. 4.2 Sampling locations along Karamana River (India) for SGI studies (Not to scale)

P3 (RP3, 22 km from P1) in the midland river, suggesting that the stream flow is fed by the groundwater in this river reach. The river stretch from P4 (RP4, 30 km from P1) to P7 (RP7, 44 km from P1) showed consistently decreasing ^{222}Rn activities (332–114 Bq/m^3) with increasing distance from upland, possibly indicating a decrease in groundwater discharge into the river in this region (Fig. 4.3).

Two notable ^{222}Rn peaks were identified in the river water samples at locations P8 (1120 Bq/m^3; RP8), i.e., 48 km from P1 and P11 (740 Bq/m^3; RP11), i.e., 56 km from P1, respectively, demonstrating the occurrence of local groundwater

Table 4.1 Indicative guide for interpreting surface water (river water) ^{222}Rn activities with regard to groundwater input (Harrington et al. 2012)

^{222}Rn (Bq/m^3)	Indicative groundwater input
<100	Negligible
100–500	Low
500–1000	Moderate
>1000	High

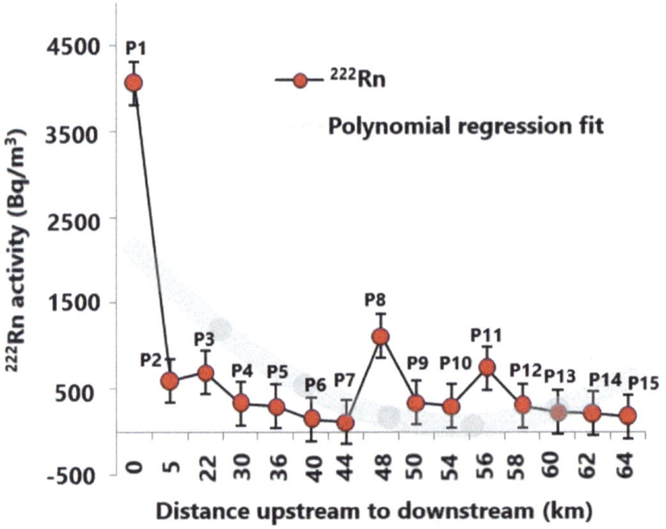

Fig. 4.3 A ^{222}Rn activities of river water samples in the profile locations (P1–P15) of Karamana River during non-monsoon (NON). *Inset* the shaded interval is a local polynomial regression (LOESS) fit

discharge into the river in these stretches. ^{222}Rn activities were found to be decreasing (Table 4.2) towards the lower reaches of the river (P12–P15; 58–64 km).

^{222}Rn activities in profile or river zone groundwater (50–150 m from river) in upstream profiles P1 and P2 of KRB (P1-0 km; P2-5 km from P1) ranged from 2810 to 63,800 Bq/m^3 (mean = 14,195.14 Bq/m^3) during non-monsoon (Fig. 4.4; Table 4.2). Groundwater ^{222}Rn activities along midstream (P3–P8; 22–48 km from P1) were generally lower compared to the upper reaches, ranging from 412 to 13,789 Bq/m^3 (mean = 4494.7 Bq/m^3). Mean ^{222}Rn activities of profile groundwater in the downstream part of river (P9–P15; 50–64 km from P1) were the lowest (Table 4.2) compared to the upper and middle river reaches, ranging from 112 to 29,567 Bq/m^3 (mean = 3302.2 Bq/m^3).

^{222}Rn activities along each profile in the river showed that most of the groundwater–river water exchange processes were confined within the near river zone groundwater (0–50 m from river). Meticulous observation of ^{222}Rn activities in groundwater also implied aquifer-aquifer interaction in most of the near (0–50 m

Table 4.2 Spatio-temporal variation of ^{222}Rn activities in the river water samples of various profiles (P1–P15) along the Karamana River course

Sr. no	Profile no	River Sample ID	Distance from P1 (km)	^{222}Rn activity (Bq/m^3)	
				Non-monsoon	Monsoon
1	P1	RP1	0	4060	1067
2	P2	RP2	5	592	442
3	P3	RP3	22	694	552
4	P4	RP4	30	332	287
5	P5	RP5	36	299	299
6	P6	RP6	40	148	126
7	P7	RP7	44	114	118
8	P8	RP8	48	1120	501
9	P9	RP9	50	344	299
10	P10	RP10	54	299	276
11	P11	RP11	56	740	268
12	P12	RP12	58	308	233
13	P13	RP13	60	229	187
14	P14	RP14	62	211	166
15	P15	RP15	64	178	143

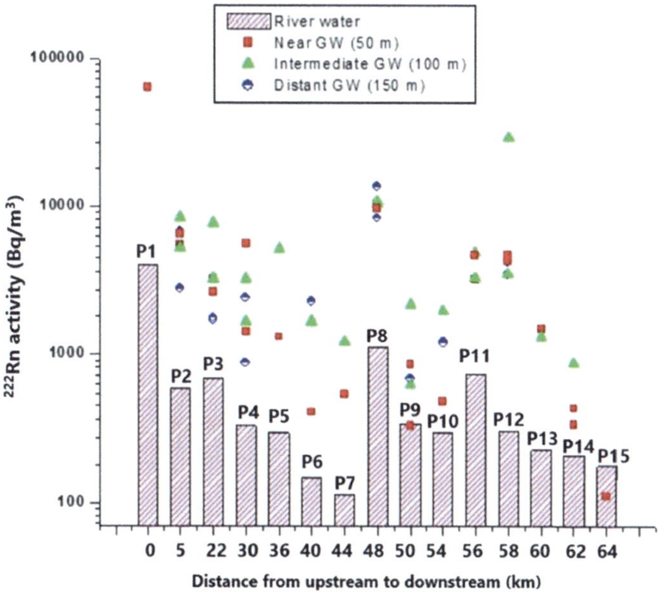

Fig. 4.4 Profile locations (P1–P15) based ^{222}Rn activities in the river water and profile (river zone) groundwater along the Karamana River during non-monsoon

from river) and intermediate (within 50–100 m from river) river zone groundwater. From Fig. 4.4, it could be inferred that P1 (0 km, highland), P2 (5 km from P1, highland), P3 (22 km, midland), P8 (48 km, midland), and P11 (56 km, lowland) are locations where river gains from groundwater contribution during non-monsoon.

(ii) **Monsoon**

In the Karamana River, a distinct decrease in ^{222}Rn activity was observed during monsoon compared to non-monsoon, owing to the dilution effect of notable rainwater input in the river basin. The ^{222}Rn activity of the river sample in the uppermost profile P1 (0 km, RP1) in highland was maximum (=1067 Bq/m^3) during monsoon. Further, it plunged (= 442 Bq/m^3; RP2) at P2 (5 km from P1) (Fig. 4.5). But, there was a slight increase in river water Rn activity (552 Bq/m^3) at P3 (22 km from P1) in the midland region.

Stored groundwater can be released rapidly to streams due to rainfall events, given the connectivity between hill slopes and river zone areas (McGuire and McDonnell 2010). River flow in this site slightly benefits from the newly rain-recharged groundwater from the near river zone aquifer (~50 m away from the river). The river stretch from P4 (30 km from P1) to P7 (44 km from P1) showed consistently decreasing ^{222}Rn activities (287–118 Bq/m^3), possibly indicating a decrease in groundwater discharge into the river in this region. However, ^{222}Rn peak (501 Bq/m^3) at P8 (48 km from P1) shows the signature of rain-sourced groundwater. ^{222}Rn activities were found to be decreasing (Fig. 4.5) further towards the lower reaches of the river (RP9–RP15; 50–64 km), suggesting the prevalence of "losing stream" during monsoon.

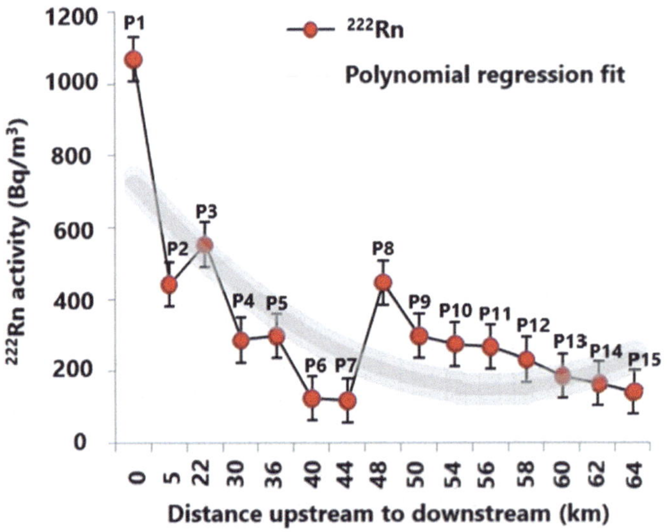

Fig. 4.5 A ^{222}Rn activities of river water samples in the profile locations (P1–P15) of the Karamana River during monsoon, KRB. *Inset* the shaded interval is a local polynomial regression (LOESS) fit

During monsoon, the upstream (P1) ^{222}Rn activities in profile (river zone) groundwater (50–150 m from river) ranged from 1650 to 26,450 Bq/m^3 (mean = 5716.4 Bq/m^3). Along the midstream reaches, represented by P3–P8 (22–48 km from P1), the groundwater ^{222}Rn activities (Table 4.2) were generally lower compared to the upper reaches, ranging from 118 to 11,466 Bq/m^3 (mean = 2685.04 Bq/m^3). Again, ^{222}Rn activities were lowest (range = 114 to 1255 Bq/m^3; mean = 954.4 Bq/m^3) in the downstream stretch compared to the upper and middle river reaches.

The observed river water–groundwater exchange processes were dominant near the river zone groundwater (0–50 m from the river) as in the non-monsoon season. In the downstream stretch from P9 to P15 (50–64 km from P1), about 71% of near river zone groundwater carried ^{222}Rn signatures close to river water, implying a higher amount of recharge from the river along this stretch. At P1 (0 km, highland), P3 (22 km, midland), and P8 (48 km, midland) profile locations, ^{222}Rn signature suggests groundwater input into the river (i.e., gaining stream) during monsoon (Fig. 4.6).

During monsoon, the upstream (P1) ^{222}Rn activities in profile (river zone) groundwater (50–150 m from river) ranged from 1650 to 26,450 Bq/m^3 (mean = 5716.4 Bq/m^3). Along the midstream reaches, represented by P3–P8 (22–48 km from P1), the groundwater ^{222}Rn activities (Table 4.2) were generally lower compared to the upper reaches, ranging from 118 to 11,466 Bq/m^3 (mean = 2685.04 Bq/m^3). Again, ^{222}Rn activities were the lowest (range = 114–1255 Bq/m^3; mean =

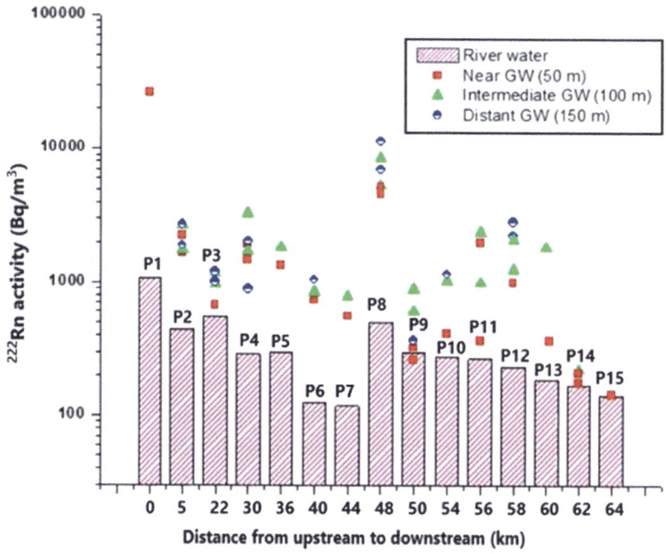

Fig. 4.6 Profile locations (P1–P15)-based ^{222}Rn activities in the river water and profile (river zone) groundwater along the Karamana River during monsoon (MON)

954.4 Bq/m^3) in the downstream stretch compared to the upper and middle river reaches.

4.10 Limitations of ^{222}Rn in SGI Studies

Although ^{222}Rn is a potential tracer in delineating SGI within a hydrological system, studies reveal that assumptions of homogenous ^{222}Rn production in alluvial systems may lead to uncertainties in estimating surface water–groundwater exchange fluxes. Peel et al. (2022) observed that ^{222}Rn emanation rates from sediments are highest within the first few meters below the surface. This uncertainty could be avoided by a thorough understanding of spatial heterogeneity of ^{222}Rn production rates, instead of assuming constant production rates. Underestimation of subsurface flow using radon tracer is possible in the case of surface water bodies connected with highly dynamic coastal systems where seawater recirculation is observed. Sadat-Noori et al. (2021) recommended the use of multi-tracers in such scenarios to reduce uncertainties.

4.11 Conclusion and Recommendations

Radon (^{222}Rn) can be used as a tracer to determine the location and magnitude of groundwater contribution to streams due to the fact that groundwater is having higher radon activity concentration compared to surface water. For obtaining the best results from this tracer approach, routine monitoring of sampling strategies (methodology, locations, frequency, etc., should be revised at local/regional/national levels with objectives to standardize the methodology.

This tracer performs well in regional scale surface water–groundwater interactions studies without interfering the system. Other than hydrological studies, ecological aspects should be taken into account while investigating surface water–groundwater interaction in hydrogeological system. Interdisciplinary approach by compiling geological, hydrological, and ecological data coupled with machine learning algorithms provides a better understanding of exchange processes between surface and groundwater entities.

In highly dynamic environments, multi-tracer approach along with field-based modeling tools should be incorporated with radon tracer method for better interpretation and estimation of hydrological fluxes. Future studies should focus on developing more sensitive and energy-efficient radon detectors for continuous measurements.

References

Abdelhalim A, Sefelnasr A, Ismail E (2019) Numerical modeling technique for groundwater management in Samalut city, Minia Governorate, Egypt. Arab J Geosci 12:1–18

Adyasari D, Dimova NT, Dulai H, Gilfedder BS, Cartwright I, McKenzie T, Fuleky P (2023) Radon-222 as a groundwater discharge tracer to surface waters. Earth-Sci Rev 104321

Appels WM, Bogaart PW, van der Zee SE (2011) Influence of spatial variations of microtopography and infiltration on surface runoff and field scale hydrological connectivity. Adv Water Resour 34(2):303–313

Balbarini N, Boon WM, Nicolajsen E, Nordbotten JM, Bjerg PL, Binning PJ (2017) A 3-D numerical model of the influence of meanders on groundwater discharge to a gaining stream in an unconfined sandy aquifer. J Hydrol 552:168–181

Bertrand G, Siergieiev D, Ala-Aho P, Rossi PM (2014) Environmental tracers and indicators bringing together groundwater, surface water and groundwater-dependent ecosystems: importance of scale in choosing relevant tools. Environ Earth Sci 72(3):813–827

Biehler A, Chaillou G, Buffin-Bélanger T, Baudron P (2020) Hydrological connectivity in the aquifer–river continuum: Impact of river stages on the geochemistry of groundwater floodplains. J Hydrol 590:125379

Binley A, Hubbard SS, Huisman JA, Revil A, Robinson DA, Singha K, Slater LD (2015) The emergence of hydrogeophysics for improved understanding of subsurface processes over multiple scales. Water Resour Res 51(6):3837–3866

Birkel C, Soulsby C (2015) Advancing tracer-aided rainfall–runoff modelling: a review of progress, problems and unrealised potential. Hydrol Process 29(25):5227–5240

Boester U, Rüde TR (2020) Utilize gadolinium as environmental tracer for surface water-groundwater interaction in Karst. J Contam Hydrol 235:103710

Carvalho FP, Oliveira JM, Malta MELM, Lemos ME (2014) Radioanalytical assessment of environmental contamination around non-remediated uranium mining legacy site and radium mobility. J Radioanal Nucl Chem 299(1):119–125

Cassidy R, Comte JC, Nitsche J, Wilson C, Flynn R, Ofterdinger U (2014) Combining multi-scale geophysical techniques for robust hydro-structural characterisation in catchments underlain by hard rock in post-glacial regions. J Hydrol 517:715–731

Chen X, Chen X (2003) Stream water infiltration, bank storage, and storage zone changes due to stream-stage fluctuations. J Hydrol 280(1–4):246–264

Cook PG (2013) Estimating groundwater discharge to rivers from river chemistry surveys. Hydrol Process 27(25):3694–3707

Cook PG, Rodellas V, Andrisoa A, Stieglitz TC (2018) Exchange across the sediment-water interface quantified from porewater radon profiles. J Hydrol 559:873–883

Ding B, Zhang Y, Yu X, Jia G, Wang Y, Wang Y, Zheng P, Li Z (2022) Effects of forest cover type and ratio changes on runoff and its components. Int Soil Water Conserv Res 10(3):445–456

Dunn SM, Bacon JR, Soulsby C, Tetzlaff D, Stutter MI, Waldron S, Malcolm IA (2008) Interpretation of homogeneity in δ18O signatures of stream water in a nested sub-catchment system in north-east Scotland. Hydrol Process Int J 22(24):4767–4782

Earon R, Riml J, Wu L, Olofsson B (2020) Insight into the influence of local streambed heterogeneity on hyporheic-zone flow characteristics. Hydrogeol J 28(8):2697–2712

Flipo N, Mouhri A, Labarthe B, Biancamaria S, Rivière A, Weill P (2014) Continental hydrosystem modelling: the concept of nested stream–aquifer interfaces. Hydrol Earth Syst Sci 18(8):3121–3149

Foster SB, Allen DM (2015) Groundwater—surface water interactions in a mountain-to-coast watershed: effects of climate change and human stressors. Adv Meteorol

Frei S, Durejka S, Le Lay H, Thomas Z, Gilfedder BS (2019) Quantification of hyporheic nitrate removal at the reach scale: exposure times versus residence times. Water Resour Res 55(11):9808–9825

Fuchs EH, King JP, Carroll KC (2019) Quantifying disconnection of groundwater from managed-ephemeral surface water during drought and conjunctive agricultural use. Water Resour Res 55(7):5871–5890

Gleeson J, Santos IR, Maher DT, Golsby-Smith L (2013) Groundwater–surface water exchange in a mangrove tidal creek: evidence from natural geochemical tracers and implications for nutrient budgets. Mar Chem 156:27–37

Haaf E, Giese M, Heudorfer B, Stahl K, Barthel R (2020) Physiographic and climatic controls on regional groundwater dynamics. Water Resourc Res 56(10):e2019WR026545

Hancock PJ, Boulton AJ, Humphreys WF (2005) Aquifers and hyporheic zones: towards an ecological understanding of groundwater. Hydrogeol J 13(1):98–111

Harrington N, Noordijn S, Cook P (2012) Evaluation of approaches to modelling surface water-groundwater interactions around drains in the South East of South Australia. Phase 1. Goyder Inst Water Res Tech Rep Ser 12(1)

Hintze S, Glauser G, Hunkeler D (2020) Influence of surface water–groundwater interactions on the spatial distribution of pesticide metabolites in groundwater. Sci Total Environ 733:139109

Holmes T, Stadnyk TA, Kim SJ, Asadzadeh M (2020) Regional calibration with isotope tracers using a spatially distributed model: a comparison of methods. Water Resourc Res 56(9):e2020WR027447

Huang P, Chui TFM (2022) Hyporheic exchange in a meandering pool-riffle stream. Water Resourc Res 58(9):e2021WR031418

Janos D, Molson J, Lefebvre R (2018) Regional groundwater flow dynamics and residence times in Chaudière-Appalaches, Québec, Canada: insights from numerical simulations. Can Water Resourc J/revue Canadienne Des Ressources Hydriques 43(2):214–239

Jiang Q, Jin G, Tang H, Shen C, Cheraghi M, Xu J, Li L, Barry DA (2020) Density-dependent solute transport in a layered hyporheic zone. Adv Water Resour 142:103645

Jin G, Yuan H, Zhang G, Zhang Z, Chen C, Tang H, Li L (2022) Effects of bed geometric characteristics on hyporheic exchange. J Hydro-Environ Res

Joo J, Tian Y, Zheng C, Zheng Y, Sun Z, Zhang A, Chang H (2018) An integrated modeling approach to study the surface water-groundwater interactions and influence of temporal damping effects on the hydrological cycle in the Miho catchment in South Korea. Water 10(11):1529

Koehn WJ, Tucker-Kulesza SE, Steward DR (2019) Conceptualizing groundwater-surface water interactions within the Ogallala aquifer region using electrical resistivity imaging. J Environ Eng Geophys 24(2):185–199

Korbel KL, Rutlidge H, Hose GC, Eberhard SM, Andersen MS (2022) Dynamics of microbiotic patterns reveal surface water groundwater interactions in intermittent and perennial streams. Sci Total Environ 811:152380

Laube G, Schmidt C, Fleckenstein JH (2022) A systematic model-based evaluation of the influence of hydraulic conductivity, heterogeneity and domain depth on hyporheic nutrient transformation. Adv Water Resour 159:104087

Leibundgut C, Maloszewski P, Külls C (2009) Tracers in hydrology. Wiley-Blackwell, Chichester, p 432

Lerner DN (2020) Groundwater recharge. In: Geochemical processes, weathering and groundwater recharge in catchments. CRC Press, pp 109–150

Li C, Gao X, Wang W, Zhang X, Zhang X, Jiang C, Wang Y (2021) Hydro-biogeochemical processes of surface water leakage into groundwater in large scale karst water system: a case study at Jinci, northern China. J Hydrol 596:125691

Magliozzi C, Grabowski RC, Packman AI, Krause S (2018) Toward a conceptual framework of hyporheic exchange across spatial scales. Hydrol Earth Syst Sci 22(12):6163–6185

Martinez JL, Raiber M, Cox ME (2015) Assessment of groundwater–surface water interaction using long-term hydrochemical data and isotope hydrology: headwaters of the Condamine River, Southeast Queensland, Australia. Sci Total Environ 536:499–516

McGuire KJ, McDonnell JJ (2010) Hydrological connectivity of hillslopes and streams: characteristic time scales and nonlinearities. Water Resourc Res 46(10)

McLachlan PJ, Chambers JE, Uhlemann SS, Binley A (2017) Geophysical characterisation of the groundwater–surface water interface. Adv Water Resour 109:302–319
Meinzer OE (1923) The occurrence of ground water in the United States with a discussion of principles, vol 50, no 846. University of Chicago
Mojarrad BB, Riml J, Wörman A, Laudon H (2019) Fragmentation of the hyporheic zone due to regional groundwater circulation. Water Resour Res 55(2):1242–1262
Mojarrad BB, Wörman A, Riml J, Xu S (2021) The influence of hyporheic fluxes on regional groundwater discharge zones. Hydrol Earth Syst Sci Discuss 1–28
Naganna SR, Deka PC, Ch S, Hansen WF (2017) Factors influencing streambed hydraulic conductivity and their implications on stream–aquifer interaction: a conceptual review. Environ Sci Pollut Res 24(32):24765–24789
Palumbo-Roe B, Wragg J, Banks VJ (2012) Lead mobilisation in the hyporheic zone and river bank sediments of a contaminated stream: contribution to diffuse pollution. J Soils Sediments 12(10):1633–1640
Peel M, Kipfer R, Hunkeler D, Brunner P (2022) Variable ^{222}Rn emanation rates in an alluvial aquifer: limits on using ^{222}Rn as a tracer of surface water–groundwater interactions. Chem Geol 599:120829
Pescimoro E, Boano F, Sawyer AH, Soltanian MR (2019) Modeling influence of sediment heterogeneity on nutrient cycling in streambeds. Water Resour Res 55(5):4082–4095
Quichimbo EA, Singer MB, Cuthbert MO (2020) Characterising groundwater–surface water interactions in idealised ephemeral stream systems. Hydrol Process 34(18):3792–3806
Rau GC, Halloran LJ, Cuthbert MO, Andersen MS, Acworth RI, Tellam JH (2017) Characterising the dynamics of surface water-groundwater interactions in intermittent and ephemeral streams using streambed thermal signatures. Adv Water Resour 107:354–369
Rivière A, Goncalves J, Jost A, Font M (2014) Experimental and numerical assessment of transient stream–aquifer exchange during disconnection. J Hydrol 517:574–583
Robinson K, Robinson CE, Roy JW, Vissers M, Almpanis A, Schneidewind U, Power C (2022) Improved interpretation of groundwater-surface water interactions along a stream reach using 3D high-resolution combined DC resistivity and induced polarization (DC-IP) geoelectrical imaging. J Hydrol 613:128468
Saedi J, Sharifi MR, Saremi A, Babazadeh H (2022) Assessing the impact of climate change and human activity on streamflow in a semiarid basin using precipitation and baseflow analysis. Sci Rep 12(1):9228
Safeeq M, Fares A (2016) Groundwater and surface water interactions in relation to natural and anthropogenic environmental changes. In: Emerging issues in groundwater resources. Springer, Cham, pp 289–326
Sahoo S, Sahoo B (2019) A geomorphology-based integrated stream–aquifer interaction model for semi-gauged catchments. Hydrol Process 33(9):1362–1377
Schubert M, Siebert C, Knoeller K, Roediger T, Schmidt A, Gilfedder B (2020) Investigating groundwater discharge into a major river under low flow conditions based on a radon mass balance supported by tritium data. Water 12(10):2838
Singh T, Gupta S, Chiogna G, Krause S, Wohlmuth B (2022) Impacts of peak-flow events on hyporheic denitrification potential. Water Resourc Res 58(3):e2021WR031407
Stefania GA, Rotiroti M, Fumagalli L, Simonetto F, Capodaglio P, Zanotti C, Bonomi T (2018) Modeling groundwater/surface-water interactions in an Alpine valley (the Aosta Plain, NW Italy): the effect of groundwater abstraction on surface-water resources. Hydrogeol J 26(1):147–162
Sun C, Wang S, Chen W (2022) Hydrochemical characteristics and the relationship between surface and groundwater in a typical 'Mountain–Oasis' ecosystem in Central Asia. Sustainability 14(12):7453
Tonina D (2012) Surface water and streambed sediment interaction: the hyporheic exchange. Fluid Mech Environ Interfaces 255–294

Trauth N, Fleckenstein JH (2017) Single discharge events increase reactive efficiency of the hyporheic zone. Water Resour Res 53(1):779–798

Trauth N, Schmidt C, Vieweg M, Maier U, Fleckenstein JH (2014) Hyporheic transport and biogeochemical reactions in pool-riffle systems under varying ambient groundwater flow conditions. J Geophys Res Biogeosci 119(5):910–928

Wang P, Yu J, Pozdniakov SP, Grinevsky SO, Liu C (2014) Shallow groundwater dynamics and its driving forces in extremely arid areas: a case study of the lower Heihe River in northwestern China. Hydrol Process 28(3):1539–1553

Wang C, Gomez-Velez JD, Wilson JL (2022) Dynamic coevolution of baseflow and multiscale groundwater flow system during prolonged droughts. J Hydrol 609:127657

Winter TC, Harvey JW, Franke OL, Alley WM (1998) Ground water and surface water—a single resource. US Geological Survey Circular 1139, 79 p. Accessed 10 Oct 2022

Wu L, Gomez-Velez JD, Krause S, Singh T, Wörman A, Lewandowski J (2020) Impact of flow alteration and temperature variability on hyporheic exchange. Water Resourc Res 56(3):e2019WR026225

Xian Y, Jin M, Zhan H, Liang X (2021) On river-aquifer exchange flow with irregular and semipervious bank. Water Resourc Res 57(10):e2020WR028984

Xiao K, Li H, Wilson AM, Xia Y, Wan L, Zheng C, Ma Q, Wang C, Wang X, Jiang X (2017) Tidal groundwater flow and its ecological effects in a brackish marsh at the mouth of a large sub-tropical river. J Hydrol 555:198–212

Yu MCL, Cartwright I, Braden JL, De Bree ST (2013) Examining the spatial and temporal variation of groundwater inflows to a valley-to-floodplain river using ^{222} Rn, geochemistry and river discharge: the Ovens River, southeast Australia. Hydrol Earth Syst Sci 17(12):4907–4924

Yuan R, Wang M, Wang S, Song X (2020) Water transfer imposes hydrochemical impacts on groundwater by altering the interaction of groundwater and surface water. J Hydrol 583:124617

Zhu A, Yang Z, Liang Z, Gao L, Li R, Hou L, Li S, Xie Z, Wu Y, Chen J, Cao L (2020) Integrating hydrochemical and biological approaches to investigate the surface water and groundwater interactions in the hyporheic zone of the Liuxi River basin, southern China. J Hydrol 583:124622

Chapter 5
Radon in Hydrograph Separation and Water Balance Studies

5.1 Hydrograph Separation

A hydrograph is a plot (Fig. 5.1) depicting the rate of flow (discharge) versus time past a specific point in a river, channel, or conduit carrying flow (Beven et al. 2015; Gore and Banning 2017). Hydrograph separation is a technique where the hydrograph is separated into its different components (baseflow and direct runoff components) to analyze flow contributions (Miller et al. 2014; Bayou et al. 2021). A better understanding of streamflow generation processes as well as catchment function is important for improved water resources management (Janssen and Ameli 2021; Li and Ameli 2022). The hydrograph separation technique using naturally occurring stable isotope tracers, viz., ^{222}Rn, δ^{18}O, and δD in which different runoff components are quantified according to their isotopic signature, is a widely used method for investigating streamflow generation processes in a river basin (Miller et al. 2021; Sukanya et al. 2022).

Depending on the physical complexity, hydrograph separation and baseflow assessment methods can be categorized into three groups (Pelletier and Andréassian 2020): (i) numerical/empirical methods, (ii) conceptual methods, and (iii) physical/chemical methods based on physico-chemical gradients (Bhaskar and Welty 2015; Meaurio et al. 2015; Birkel and Soulsby 2015). The following sections illustrate a quick picture of these three categories.

5.2 Empirical and Numerical Methods

Numerous traditional baseflow-separation techniques rely on processing the hydrograph as a signal rather than hydrological factors. The majority of these techniques are predicated on the idea that surface runoff has a substantially shorter transfer time than groundwater and that this duration is generally constant between precipitation events. These earliest techniques were graphical. A straight line is drawn from the

S. S. and S. Joseph, *Environmental Radon*, Environmental Science and Engineering,
https://doi.org/10.1007/978-981-99-2672-5_5

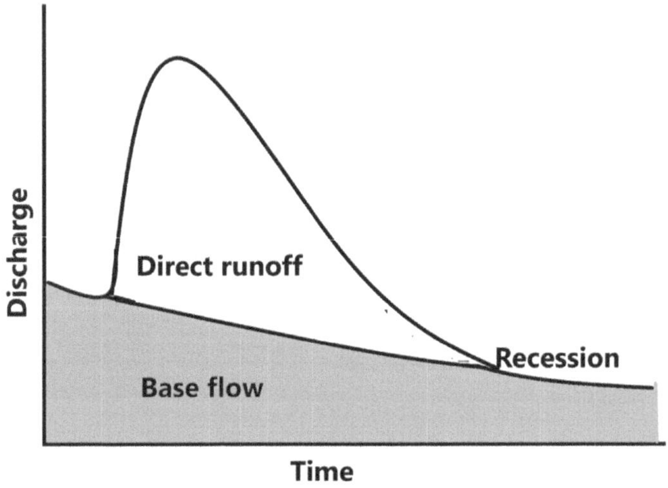

Fig. 5.1 Basic flow components of hydrograph (*adapted from* Schroeder 2003)

start of each peak to the point N days, after recognizing the peaks along the hydrograph and determining the surface runoff time constant—"n" number of days (Foks et al. 2019; Pelletier and Andréassian 2020). This direct method has been refined to account for succeeding precipitation occurrences or aquifer recharging during rainfall, but it is still very arbitrary due to the difficulty in automatization.

In a hydrograph, streamflow component beneath the dividing line is assumed to be baseflow, whereas the remaining peaks are considered as surface flow component. Using this hypothesis, low-pass numerical filtering of hydrograph has been applied as a baseflow-separation method. Alike electronic signal filtering technique, the highest signal frequencies (noise) are regarded as outliers and eliminated, while low frequencies are kept for making the time series smoother. The Fourier transform-based filters function based on this configuration and are used to smooth the data time series. Even quick components of streamflow have low-frequency Fourier terms.

Recently, Solgi et al. (2022) introduced a new wavelet transform (WT) method to separate baseflow discharge of the Gamasiyab karstic spring (Western Iran). Artificial neural networks have been employed as data-driven models for hydrograph separation (Tongal and Booij 2022). Numerical methods require one or multiple parameters (e.g., linear filter coefficients, time interval width, etc.) prior processing (Mohammadi et al. 2019; Pelletier and Andréassian 2020; Pozdniakov et al. 2022).

5.3 Conceptual Methods

Although numerical methods are convenient and simple options for hydrograph separation, they often fail to explain subsurface hydrology without uncertainties. This limitation could be overcome by using conceptual models where aquifers are represented as conceptual linear reservoirs with overland flow, interflow, and baseflow components. Here input is assumed to be zero, thus full recession curve between rainfall events is generated (Schmidt et al. 2014). Casado-Rodríguez and del Jesus (2022) analyzed the equifinality in the calibration of a conceptual model by calibrating outflows of the different runoff reservoirs against a time-series representative of the process.

5.4 Physico-chemical Methods

Physico-chemical methods are based on the distinct signatures of baseflow and surface flow components contributing to the total flow (Buttle 1994; Cartwright 2022). The known values of the parameters could be applied in a simple mass balance approach to delineate the fraction of each contributing component. The most common method applied for this purpose is tracer-based baseflow-separation technique. A tracer is selected based on its signature difference in each hydrological component—groundwater, river water, rainwater, etc.

The commonly used tracers for this purpose are isotope tracers—radon (^{222}Rn), deuterium (^{2}H or D), or oxygen-18 (^{18}O); ions (carbonates, calcium, magnesium, sulfates) from dissolved minerals or silica; and the known concentrations of these tracers are applied in mass balance assessment (Sukanya et al. 2022; Shao et al. 2020; De Filippi et al. 2021) which allows for a more realistic dynamic baseflow decomposition. Other methods rely on conductivity, pH, or temperature, which can be easier to measure in situ and real-time monitoring.

Kirchner (2019) proposed a statistical tracer-based separation to replace this mass balance hypothesis using a regression between baseflow and tracer concentrations in streamflow and rainfall. Although tracer-based approaches can offer relevant insights on the hydrological processes within a basin, regular monitoring of these tracer concentrations is not applicable in the case of most catchments.

5.5 Radon Tracer for Hydrograph Separation

The basis of isotope hydrograph separation is the concept that the isotopic compositions of relevant runoff compartments are distinct. This isotopic signal within each compartment is constant over both space and time (Penna and van Meerveld 2019).

Along a river, radon in groundwater is collected assuming that the river and groundwater components are hydraulically connected (Hofmann 2023). In some other studies, the pore water radon activity concentration in the river can be estimated based on ^{226}Ra concentrations in the sediment matrix (Khadka et al. 2017).

Similar to the aquifer, ^{226}Ra decay constantly produces ^{222}Rn in the river sediment. The radon concentration in the sediment pore water can be utilized as a proxy for the radon concentration in the groundwater since the local aquifer host rock controls the river sediments. River bed sediment samples from each river segment should be collected for this purpose. From upstream to downstream of the river segment under investigation, samples should be taken from each profile that traverses the river.

5.6 Radon Loss by Degassing

Radon degassing is one of the important sinks used in radon mass balance approach. Degassing is often primarily fueled by stream water turbulence (Schubert et al. 2020a, b). The loss of gases like radon to the atmosphere from a stream or any other body of surface water can be measured using a variety of techniques. The techniques include injecting tracer gases like propane or SF_6 (Gleeson et al. 2018; Jensen et al. 2022), measuring the concentration of gas in river reaches that are losing gas (Cartwright et al. 2011), and currently most frequently using empirical equations that relate the physical characteristics of the river and the flow regime to the gas loss (Atkinson et al. 2015). All of these techniques produce a degassing coefficient, k (d^{-1}), which is used to calculate the amount of gas (here referred to as radon) lost from the water body.

Schubert et al. (2020a, b) applied an empirical equation for quantifying the degassing rate using the length scale and the river discharge rate. Empirical equations were initially developed to quantify re-aeration of streams (Avery et al. 2018). Gas transfer rates (k) were estimated using the O'Connor and Dobins (1958) and Negulescu and Rojanski (1969) gas transfer models as modified by Genereux and Hemond (1992) and Mullinger et al. (2007):

$$k = 9.301 \times 10^{-3}\left(\frac{v^{0.5}}{D^{1.5}}\right) \tag{5.1}$$

Unland et al. (2013) used an additional equation:

$$k = 4.87 \times 10^{-4}\left(\frac{v}{d}\right)^{0.85} \tag{5.2}$$

where d is river depth (m) and v is stream velocity (mday^{-1}) as calculated from discharge, depth, and width data.

Due to distinct levels of radon emanation in various media, it is a potential tracer when combined with other chemical tracers like electrical conductivity, chemical ions, etc. to distinguish the components of superficial water (vadose zone water), soil groundwater, and bedrock water in the total river flow (Genereux and Hemond 1992). Aquifer-supplied groundwater is not the only component of typical baseflow. Water from the vadose zone or the soil makes up a portion of baseflow.

It is crucial to comprehend the relative fluxes of both groundwater and soil water and how their summated values determine baseflow, according to earlier studies utilizing ^{222}Rn as a tracer. There is a distinct difference between vadose zone water (relatively lower ^{222}Rn) and soil groundwater (higher ^{222}Rn), especially in a groundwater-fed river basin, and the radioactive decay constant of ^{222}Rn (0.181 day^{-1}) in which vadose zone water attains the ^{222}Rn signature of soil groundwater makes it suitable for hydrograph separation.

In a micro-basin of the Attert River in Luxembourg, Kies et al. (2005) measured radon along with electrical conductivity and discharge rate. Then, a two-component mass balance equation was used to separate the contributions of surface water and groundwater fluxes to the river. The measured groundwater radon activities were found to be about tenfold higher than that in river. As the water trapped in underground fracture zones is pushed into flow channels, every significant rain event resulted in a rapid drop in electric conductivity and a rise in radon activity.

Wan et al. (2019) developed isotope mass models combining tritium and ^{222}Rn measurements to quantify the influence of permafrost degradation on surface and subsurface water systems in the source region of Yellow River, Qinghai–Tibet Plateau (China). The study demonstrated the spatial variability of permafrost thaw impacts and confirmed a significant influence on surface and subsurface runoff processes, including effects on water yield capacity and drainage connectivity.

5.7 Radon Tracer in Glacial Hydrological Systems

Glaciers are one of the most vulnerable victims of global warming. A thorough knowledge on changes and evolution of glaciers based on meltwater drainage within glacier systems is very important (Eyles 2006). Isotopic signatures of meltwaters are often used as important proxies for understanding the fate and evolution of meltwaters circulating in glaciers (Kies et al. 2015). As the water flows along the glacier drainage system, it becomes enriched with different ionic species and radon due to considerable contact with bedrock.

High radon activity concentrations indicate contact with radium in the matrix occurring less than 20 days prior sampling, i.e., contact duration is an important factor. In contrast, low mineralized, low radon activity concentrations imply a short period of flow along the glacier bed, via a hydraulically effective drainage pathway characterized by large water fluxes and rapid transit speeds. As radon activity concentrations depend on radium concentrations of rocks, the rock/sediment-water contact surface and the contact duration (Sukanya et al. 2021), radon in meltwater serves as a reliable indicator of water component's recent contact with bedrock.

Since noble gas isotopic tracers are chemically and physically inert, they sparsely react with catchment variables compared to other chemical tracers. This property allows the use of natural radionuclides as environmental tracers in hydrograph separation. Here, one can anticipate a contribution in separating the origin of meltwaters into three distinct categories: on the glacier's surface (supraglacial), inside (englacial), and below ice (subglacial) (Fyffe et al. 2019).

Albeit being soluble, radon tends to partition into the air phase when introduced into an air–water system. Evasion loss is a function of temperature and radon activity concentrations present in the air, hence this could result in uncertainties during late melt seasons when subglacial channels may not packed with water. Also, water flow through subglacial channels is often with high velocity and the turbulence reaches its peak during summer melt season. This results in gas exchange loss. Therefore, if evasion is the key mechanism controlling radon in the glacier outflow system, lower radon activities may be expected towards the end of a falling discharge limb of hydrograph.

Every reservoir that contributes to streamflow has common conservative tracers (such as chloride, ^{18}O), including snowmelt, precipitation, overland, soil, and groundwater. This is not the case with ^{222}Rn, which can be used in separation studies without any major interference which makes it advantageous. So it is possible to identify the sources of subterranean stream flow using ^{222}Rn concentrations. Additionally, as tracer concentrations change across time and space, it becomes difficult to divide streamflow into its relative components. Groundwater ionic concentrations depend on the flow rate, the amount of time spent there, the rate of kinetic mineral weathering, and the amount of weatherable material that is available on the surface (Xia et al. 2022). Because of this, applying average end member chemical analysis to groundwater might be difficult because groundwater chemical fingerprints can change with prolonged residence durations. As it takes about 2 weeks for radon to reach equilibrium in the subsurface, it can be assumed to be at steady state in most cases.

Usually, the conventional mass balance equations used for conservative tracers are not applicable for radioactive tracers existing in gas phase. Rather, a model that takes into account subsurface discharge, gas exchange with the environment, and radioactive decay is needed to estimate subsurface input from ^{222}Rn. Hence, Cook et al. (2006) proposed a 1D stream transport model for simulating longitudinal radon activity using the following equations:

$$\frac{\partial Q}{\partial x} = I(x) - L(x) - E(x) \tag{5.3}$$

$$Q\frac{\partial Q}{\partial x} = I(ci - c) + wEc - kwc - dw\lambda c \tag{5.4}$$

where c is the activity concentration of radon in the stream (Bq/L), ci is the concentration of radon in groundwater inflow (Bq/L), I is the groundwater inflow rate (m^3/m/day), w is the stream width (m), d is the mean stream depth (i.e., cross-sectional area/width) (m), k is the gas exchange velocity across the water surface (m/day), λ is the radon decay coefficient (0.181 d^{-1}, i.e., 0.181 per day), Q is the stream discharge (m^3/day), E is the evaporation rate (m/day), L is the stream water loss rate by means of direct pumping or outflow to the aquifer (m^3/m/day), and x is distance in the direction of flow (m). Equation 5.3 is the mass balance equation for discharge in the stream, and Eq. 5.4 is the 1D, steady-state equation for solute transport in the stream.

The model assumes steady-state flow conditions, zero atmospheric concentration, and no generation of ^{222}Rn within stream (Cook et al. 2006). By altering the groundwater discharge step function after Gardner (2011), distributed groundwater discharge was calculated using a least-squares fit of the modeled and observed ^{222}Rn. Total subsurface discharge, which includes groundwater and soil water, was combined into one term for each sample period. Then, assuming snowmelt contains no radon; total subsurface discharge was used to calculate snowmelt discharge.

Fractions of subsurface discharge to total streamflow (Fss) were computed to assess incremental gains in subsurface sources at each stilling well:

$$F_{ss} = \frac{Q_{ss1} - Q_{ss1}}{\Delta Q_t} \tag{5.5}$$

where Q_{ss1} is the upstream modeled subsurface discharge (L/s), Q_{ss2} is the downstream modeled subsurface discharge (L/s), and ΔQt is the difference in measured total streamflow from upstream to downstream (L/s).

Being the major contributor of uncertainty in a model, parameters that cannot be measured in the field should receive special attention. In low-order mountainous streams with varying geometries, velocities, and temperatures, gas transfer velocity (k) is challenging to limit. The gas transfer velocity could be roughly calculated using equations based on slope, velocity, and depth (O'Connor and Dobbins 1958; Negulescu and Rojanski 1969; Raymond et al. 2012; Holgerson et al. 2017).

5.8 Reducing Uncertainties During Sampling and Measurement

It is frequently challenging to sample in situ from beneath the ice sheet. For ^{222}Rn estimation in these circumstances, laboratory-based equilibration experiments employing sediments released from subglacial environments can be used. Separate aliquots of sediment samples from the proglacial river should be incubated with Ra-free water for more than five half-lives (>20 days) in a sealed 1 L HDPE bottle. Preferably, these samples weigh 50 g each. A radon-free carrier gas (E.g., helium) should be used to flush samples into cold traps and scintillation cells before they are triple-counted and evaluated. Using wet bulk densities and porosities, wet sediment ^{222}Rn activities in dpm g^{-1} should be translated to pore water ^{222}Rn activities (dpm L^{-1}).

A possible source of ^{222}Rn subsurface is the diffusion of ^{222}Rn from sediments in subglacial channels (Linhoff et al. 2017). Therefore, the laboratory approach developed by Chanyotha et al. (2014) can be used to calculate the diffusive flux. For this procedure, an airtight reaction flask that is linked to the closed loop with RAD7 must be filled with 100 g of wet subglacial sediment and 500 mL of ^{222}Rn-free water. Then, using the integrated RAD7 pump, air is forced via a gas diffusion stone submerged in the water phase, past a desiccant, and back to the radon analyzer where the activity is detected and recorded.

Gas leakage is low for regular measurements, but, during the multi-day experiment, it is necessary to rectify a known tiny leak within the internal air pump of the RAD7 detector. The near-linear slope of ^{222}Rn activity in the reaction flask versus time over the initial several hours of the experiment is used to calculate the diffusive fluxes. The diffusive flux uncertainty is estimated using slope uncertainty.

5.9 Radon in Water Balance Studies

Radon is an important tracer in water balance studies in diverse water bodies, viz., lakes, floodplains, etc. The following sub-sections account on these applications of radon.

5.9.1 Understanding the Hydrodynamics and Water Balance of Lakes

The ecosystem services offered by lakes, such as water and sediment transport, nutrient retention, biogeochemical processing, climate change mitigation, biodiversity, and hydrological regulation, are significantly impacted by hydrodynamics and water balance (Schallenberg et al. 2013; Burnett et al. 2017; Sterner et al. 2020; Sun et al. 2021), among other factors. Lacustrine Groundwater Discharge (LGD) research is currently gaining momentum, similar to Submarine Groundwater Discharge (SGD) studies from the forthcoming Chap. 6 (Luo et al. 2018; Wallace et al. 2021). Due to the methodological challenges in estimating LGD, which is defined as the groundwater exfiltration from lakeshore aquifers to lakes, it is frequently disregarded in studies of lake water balance (Petermann et al. 2018).

To comprehend hydrodynamics and assess the lake water balance, radon measurements in the lake and its inflow components can be used (Luo et al. 2016; Wilson and Rocha 2016). The hydrodynamics of the lake and the groundwater inflow locations are both determined by the geographical and temporal distributions of radon in the lake water.

The water balance equation for a well-mixed lake can be expressed as follows under steady-state conditions:

$$Q_{si} + Q_{gi} + P \cdot A = Q_{so} + Q_{go} + E \cdot A \quad (5.6)$$

where Q_{si} is the surface water inflow (m^3/d), Q_{gi} is the groundwater inflow (m^3/d), P is the amount of rainfall over the lake (m/d), A is the area of the lake (m^2), Q_{so} is the surface outflow from the lake (m^3/d), Q_{go} is the subsurface lake water exfiltration (m^3/d), and E is the evaporation rate (m/d).

The radon mass balance equation can be written as

$$Q_{si} \cdot C_{si} + Q_{gi} \cdot C_{gi} + F_{diff} \cdot A = Q_{so} \cdot C_{Li} + Q_{go} \cdot C_{Li} + F_{atm} \cdot A + \lambda.V.C_{Li} \quad (5.7)$$

where C_{si}, C_{gi}, and C_{Li} are the ^{222}Rn activity concentrations of the inflowing surface water, groundwater, and the lake, respectively (Bq/m^3); F_{diff} is the diffusive flux of ^{222}Rn from the lake sediments (Bq/m^2d); F_{atm} is the rate of radon degassing (exchange) to the atmosphere (Bq/m^2d); λ is the decay constant of ^{222}Rn ($0.18\ d^{-1}$); and V is the volume of the lake (m^3).

Equation 5.6 assumes that under a hydrologic steady state, the sum of all inflowing waters (groundwater, surface water, and precipitation on the lake) equals the outflowing components (surface and groundwater outflow and evaporation from the lake). Similarly, the ^{222}Rn balance equation (Eq. 5.7) represents the total ^{222}Rn flux entering and leaving the lake. The groundwater inflows and subsurface outflows of a lake can be calculated by solving the water and radon mass balance equations.

Vertical gradients in ^{222}Rn can reveal important details about groundwater input sources in lakes. Since most hydrodynamic studies assumed a well-mixed water column, they ignored this vertical stratification (Isokangas et al. 2015; Arnoux et al. 2017). A multi-box model for radon dispersion in a lake was developed by Kluge et al. (2012) that takes into account the vertical inhomogeneity of groundwater inflow and transit between the boxes. They also used the depth-dependent computations to restrict a number of ^{222}Rn mass balance parameters. In consistent with this idea, Arnoux et al. (2017) observed that multilayer mass balance models outperformed single-layered well-mixed models in quantifying the temporal variations of groundwater exchange fluxes.

In order to determine LGD in a glacial lake, Wallace et al. (2021) coupled radon data with regional scale groundwater–surface water modeling. They found that successful radon-based techniques require significant inputs regarding the source and sink. Quantifying LGD rates and associated geochemical cycling requires a thorough understanding of the geological conditions (Keim et al. 2019; Sun et al. 2021). For example, preferential routes spanning from large scale (buried paleochannel) to small scale (plant roots) had a significant influence in enhancing the

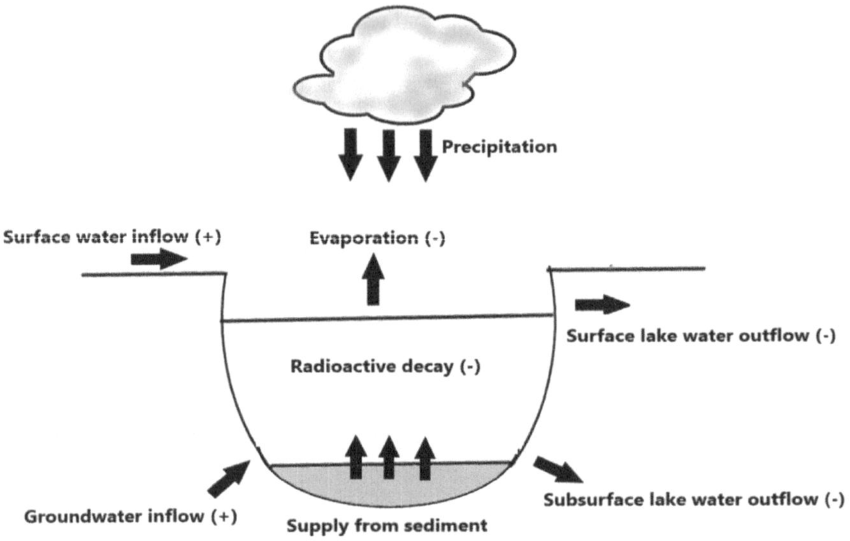

Fig. 5.2 Schematic diagram on the water balance of a lake (*Source* Sukanya et al. 2022)

aquifer permeability when Sun et al. (2021) found elevated LGD rates in a porous shoreline aquifer.

When it comes to radon in water analysis, sampling is frequently a challenging task. There is a high probability of inaccuracy while sampling lake water because of radon loss through degassing. To prevent this error, in situ radon extraction directly from lake water is recommended. This can also be done by utilizing a water pump (Wang et al. 2020). Three types of extraction units—a flow-through spray chamber, a flow-through membrane extraction module, and a submersible or coiled membrane tube—were compared by Schubert et al. (2012). Based on the observations, it was concluded that the spray chamber is apt for turbid waters whereas flow-through membrane is suitable for clear water conditions, and submersible membrane tube was found to be suitable for low-yielding groundwater with short-term radon variations (Fig. 5.2).

5.9.2 Understanding the Hydrodynamics and Water Balance in Floodplains

Floodplains are important hydrological and ecological corridors with important functions, viz., regulating runoff generation, biogeochemical transformations for natural attenuation of pollutants, etc. (Wohl 2021). Uncertainty underlies the role of flood events in the dynamics of the groundwater connected to wetlands and floodplains. However, several short-term studies have noted significant increase in groundwater discharge after flood events (de Weys et al. 2011; Gilfedder et al. 2015), which is in line with aforementioned theory.

The overall water balance of wetlands and floodplains may be significantly impacted by flood-induced groundwater pulses. Increased groundwater solute discharge has also been associated with severe acidification, deoxygenation, and high carbon dioxide levels in some rivers during floods (Santos et al. 2013; Atkins et al. 2013).

However, the dynamics of groundwater in wetlands and floodplains during flood events remains a largely unexplored area of research. Despite this, there have been a number of studies that have provided evidence of significant increase in groundwater discharge following flood events, with findings supporting the relevant theories (de Weys et al. 2011; Gilfedder et al. 2015). In a post-flood scenario, these flood-induced groundwater pulses can have significant impacts on the overall water balance of wetlands and floodplains, including the discharge of solutes and the associated severe acidification, deoxygenation, and elevated CO_2 levels in some rivers (Santos et al. 2013; Atkins et al. 2013). Thus, a comprehensive understanding of the water balance and hydrodynamic processes of floodplain systems is crucial for effective water resources management.

In their study, Williams et al. (2020) utilized ^{222}Rn as a groundwater tracer to develop a conceptual model of the hydrological function of the Mulwaree River chain-of-ponds system (SW Sydney, Australia) over an extended period of time that incorporated dry periods and large rainfall events. The ponds were found to be connected during high-flow events by preferential flow paths and a small fraction of groundwater input into the ponds from floodplain aquifer was observed based on radon signatures. The study highlighted the delicate balance of the hydrological function of this chain-of-ponds system, which was potentially sensitive to changes in climate altering rainfall and evaporation rates and local-scale groundwater interference activities emphasizing the importance of utilizing ^{222}Rn as a tracer to better understand wetland hydrological systems.

Conservative tracers, viz., stable isotopes (e.g., δ^2H and $\delta^{18}O$) and radioactive isotopes (^{222}Rn) of water are employed to differentiate "new" and "old" water in these systems with highly dynamic, complex surface water–groundwater interactions (Kendall and Caldwell 1998). Radon is sensitive to quick exchanges between surface and groundwater entities because it has a shorter half-life than other radioactive isotope tracers (e.g., tritium). This makes it an ideal tracer for separating surface and subsurface flows, which can rapidly switch direction during flood events.

Studies that integrate both the quantitative and qualitative advantage of assessing groundwater dynamics over different timescales are limited (Wilson et al. 2015). Atkins et al. (2013) performed high-resolution ^{222}Rn and pCO_2 measurements in a highly modified tidal floodplain creek and estuary to assess the groundwater contributed CO_2 in surface waters. Post-precipitation (245 mm) measurements after 2 days revealed high ^{222}Rn activities (~86.1 dpm L^{-1}) and high pCO_2 (~11,217 μatm), suggesting a significant groundwater influence after flood event. Similarly, Webb et al. (2016) applied the mass balance method to quantify groundwater discharge in an artificially agricultural floodplain. CO_2, CH_4, and groundwater dynamics were quantified using ^{222}Rn tracer during the rapid recession of flood event.

5.10 Conclusion and Outlook

Although various techniques (numerical/empirical, conceptual, physico-chemical methods) are available for separating surface runoff and baseflow components in a hydrograph, application of isotopes like radon along with these methods improves the level of interpretation. Radon is a potential tracer which could be deployed in hydrograph separation and water balance studies for a wide variety of hydrological systems including rivers, lakes, glacial systems, flood plains, etc. The development of new and improved techniques for analyzing and interpreting radon data will continue to enhance our understanding of the movement and interaction of water in various hydrological systems.

As we conclude this chapter on radon in water balance studies, it is important to note that sampling for radon analysis in water can be a challenging task, especially in lakes. However, there are methods such as in situ radon extraction from lake water that can help reduce the uncertainties associated with sampling. Furthermore, hydrodynamic models have been integrated with the radon mass balance method to reduce uncertainties in flux calculations. Nevertheless, it is important to keep in mind that the accuracy of these models is dependent on the assumptions and inputs used during their application.

References

Arnoux M, Gibert E, Barbecot F, Guillon S, Gibson J, Noret A (2017) Interactions between groundwater and seasonally ice-covered lakes: using water stable isotopes and radon-^{222} multilayer mass balance models. Hydrol Process 31(14):2566–2581

Atkins ML, Santos IR, Ruiz-Halpern S, Maher DT (2013) Carbon dioxide dynamics driven by groundwater discharge in a coastal floodplain creek. J Hydrol 493:30–42

Atkinson AP, Cartwright I, Gilfedder BS, Hofmann H, Unland NP, Cendón DI, Chisari R (2015) A multi-tracer approach to quantifying groundwater inflows to an upland river; assessing the influence of variable groundwater chemistry. Hydrol Process 29(1):1–12

Avery E, Bibby R, Visser A, Esser B, Moran J (2018) Quantification of groundwater discharge in a subalpine stream using radon-^{222}. Water 10(2):100

Bayou WT, Wohnlich S, Mohammed M, Ayenew T (2021) Application of hydrograph analysis techniques for estimating groundwater contribution in the Sor and Gebba streams of the Baro-Akobo river Basin, southwestern Ethiopia. Water 13(15):2006

Beven K, Bathurst J, O'Connell E, Littlewood I, Blackie J, Robinson M (2015) Hydrological modelling. In: Progress in modern hydrology: past, present and future. Wiley-Blackwell, pp 216–239

Bhaskar AS, Welty C (2015) Analysis of subsurface storage and streamflow generation in urban watersheds. Water Resour Res 51(3):1493–1513

Birkel C, Soulsby C (2015) Advancing tracer-aided rainfall–runoff modelling: a review of progress, problems and unrealised potential. Hydrol Process 29(25):5227–5240

Burnett WC, Wattayakorn G, Supcharoen R, Sioudom K, Kum V, Chanyotha S, Kritsananuwat R (2017) Groundwater discharge and phosphorus dynamics in a flood-pulse system: Tonle Sap Lake, Cambodia. J Hydrol 549:79–91

Buttle JM (1994) Isotope hydrograph separations and rapid delivery of pre-event water from drainage basins. Prog Phys Geogr 18(1):16–41

Cartwright I (2022) Implications of variations in stream specific conductivity for estimating baseflow using chemical mass balance and calibrated hydrograph techniques. Hydrol Earth Syst Sci 26(1):183–195

Cartwright I, Hofmann H, Sirianos MA, Weaver TR, Simmons CT (2011) Geochemical and ^{222}Rn constraints on baseflow to the Murray River, Australia, and timescales for the decay of low-salinity groundwater lenses. J Hydrol 405(3–4):333–343

Casado-Rodríguez J, del Jesus M (2022) Hydrograph separation for tackling equifinality in conceptual hydrological models. J Hydrol 610:127816

Chanyotha S, Kranrod C, Burnett WC (2014) Assessing diffusive fluxes and pore water radon activities via a single automated experiment. J Radioanal Nucl Chem 301:581–588

Cook PG, Favreau G, Dighton JC, Tickell S (2003) Determining natural groundwater influx to a tropical river using radon, chlorofluorocarbons and ionic environmental tracers. J Hydrol 277(1–2):74–88

Cook PG, Lamontagne S, Berhane D, Clark JF (2006) Quantifying groundwater discharge to Cockburn River, southeastern Australia, using dissolved gas tracers ^{222}Rn and SF6. Water Resour Res 42(10)

De Filippi FM, Iacurto S, Grelle G, Sappa G (2021) Magnesium as environmental tracer for karst spring baseflow/overflow assessment—a case study of the Pertuso karst spring (Latium Region, Italy). Water 13(1):93

de Weys J, Santos IR, Eyre BD (2011) Linking groundwater discharge to severe estuarine acidification during a flood in a modified wetland. Environ Sci Technol 45(8):3310–3316

Eyles N (2006) The role of meltwater in glacial processes. Sed Geol 190(1–4):257–268

Foks SS, Raffensperger JP, Penn CA, Driscoll JM (2019) Estimation of base flow by optimal hydrograph separation for the conterminous United States and implications for national-extent hydrologic models. Water 11(8):1629

Fyffe CL, Brock BW, Kirkbride MP, Black AR, Smiraglia C, Diolaiuti G (2019) The impact of supraglacial debris on proglacial runoff and water chemistry. J Hydrol 576:41–57

Gardner WP, Harrington GA, Solomon DK, Cook PG (2011) Using terrigenic 4He to identify and quantify regional groundwater discharge to streams. Water Resour Res 47(6)

Genereux DP, Hemond HF (1992) Determination of gas exchange rate constants for a small stream on Walker Branch Watershed, Tennessee. Water Resour Res 28(9):2365–2374

Gilfedder BS, Frei S, Hofmann H, Cartwright I (2015) Groundwater discharge to wetlands driven by storm and flood events: quantification using continuous Radon-^{222} and electrical conductivity measurements and dynamic mass-balance modelling. Geochim Cosmochim Acta 165:161–177

Gleeson T, Manning AH, Popp A, Zane M, Clark JF (2018) The suitability of using dissolved gases to determine groundwater discharge to high gradient streams. J Hydrol 557:561–572

Gore JA, Banning J (2017) Discharge measurements and streamflow analysis. In: Methods in stream ecology, vol 1. Academic Press, pp 49–70

Hofmann H (2023) Estimating the role of bank flow to stream discharge using a combination of baseflow separation and geochemistry. Water 15(5):844

Holgerson MA, Farr ER, Raymond PA (2017) Gas transfer velocities in small forested ponds. J Geophys Res Biogeosci 122(5):1011–1021

Isokangas E, Rozanski K, Rossi PM, Ronkanen AK, Kløve B (2015) Quantifying groundwater dependence of a sub-polar lake cluster in Finland using an isotope mass balance approach. Hydrol Earth Syst Sci 19(3):1247–1262

Janssen J, Ameli AA (2021) A hydrologic functional approach for improving large-sample hydrology performance in poorly gauged regions. Water Resour Res 57(9):e2021WR030263

Jensen CR, Genereux DP, Gilmore TE, Solomon DK, Mittelstet AR, Humphrey CE, MacNamara MR, Zeyrek C, Zlotnik VA (2022) Estimating groundwater mean transit time from SF6 in stream water: field example and planning metrics for a reach mass-balance approach. Hydrogeol J 30(2):479–494

Keim C, Mehler F, Wolf T, Gilfedder B (2019) Mapping spatial patterns of groundwater discharge in a deep lake using high-resolution temperature sensors. Inland Waters 9(3):334–344

Kendall C, Caldwell EA (1998) Fundamentals of isotope geochemistry. In: Isotope tracers in catchment hydrology. Elsevier, pp 51–86

Khadka MB, Martin JB, Kurz MJ (2017) Synoptic estimates of diffuse groundwater seepage to a spring-fed karst river at high spatial resolution using an automated radon measurement technique. J Hydrol 544:86–96

Kies A, Hengesch O, Tosheva Z, Nawrot AP, Jania J (2015) Overview on radon measurements in Arctic glacier waters. Cryosphere Discuss 9(2):2013–2052

Kies A, Hofmann H, Tosheva Z, Hoffmann L, Pfister L (2005) Using ^{222}Rn for hydrograph separationin a micro basin (Luxembourg). Annals of Geophys 48(1)

Kirchner BM, Tetef TJ, Dutton JC, Elliott GS (2019) Three-dimensional entrainment dynamics of a supersonic base flow. In: AIAA Scitech 2019 forum, p 0071

Kluge T, Von Rohden C, Sonntag P, Lorenz S, Wieser M, Aeschbach-Hertig W, Ilmberger J (2012) Localising and quantifying groundwater inflow into lakes using high-precision ^{222}Rn profiles. J Hydrol 450:70–81

Li H, Ameli A (2022) A statistical approach for identifying factors governing streamflow recession behaviour. Hydrol Process 36(10):e14718

Liao F, Cardenas MB, Ferencz SB, Chen X, Wang G (2021) Tracing bank storage and hyporheic exchange dynamics using ^{222}Rn: virtual and field tests and comparison with other tracers. Water Resour Res 57(5):e2020WR028960

Linhoff BS, Charette MA, Nienow PW, Wadham JL, Tedstone AJ, Cowton T (2017) Utility of ^{222}Rn as a passive tracer of subglacial distributed system drainage. Earth Planet Sci Lett 462:180–8.

Luo X, Jiao JJ, Wang XS, Liu K (2016) Temporal ^{222}Rn distributions to reveal groundwater discharge into desert lakes: Implication of water balance in the Badain Jaran Desert, China. J Hydrol 534:87–103

Luo X, Kuang X, Jiao JJ, Liang S, Mao R, Zhang X, Li H (2018) Evaluation of lacustrine groundwater discharge, hydrologic partitioning, and nutrient budgets in a proglacial lake in the Qinghai-Tibet Plateau: using ^{222} Rn and stable isotopes. Hydrol Earth Syst Sci 22(10):5579–5598

Meaurio M, Zabaleta A, Uriarte JA, Srinivasan R, Antigüedad I (2015) Evaluation of SWAT models performance to simulate streamflow spatial origin. The case of a small forested watershed. J Hydrol 525:326–334

Miller MP, Susong DD, Shope CL, Heilweil VM, Stolp BJ (2014) Continuous estimation of baseflow in snowmelt-dominated streams and rivers in the Upper Colorado R iver B asin: a chemical hydrograph separation approach. Water Resour Res 50(8):6986–6999

Miller SA, Mercer JJ, Lyon SW, Williams DG, Miller SN (2021) Stable isotopes of water and specific conductance reveal complimentary information on streamflow generation in snowmelt-dominated, seasonally arid watersheds. J Hydrol 596:126075

Mohammadi F, FakheriFard A, Ghorbani MA (2019) Application of cross-wavelet–linear programming–Kalman filter and GIUH methods in rainfall–runoff modelling. Environ Earth Sci 78(5):1–20

Mullinger NJ, Binley AM, Pates JM, Crook NP (2007) Radon in Chalk streams: spatial and temporal variation of groundwater sources in the Pang and Lambourn catchments, UK. J Hydrol 339(3–4):172–182

Negulescu M, Rojanski V (1969) Recent research to determine reaeration coefficient. Water Res 3(3):189–202

O'Connor DJ, Dobbins WE (1958) Mechanism of reaeration in natural streams. Trans Am Soc Civ Eng 123(1):641–666

Pelletier A, Andréassian V (2020) Hydrograph separation: an impartial parametrisation for an imperfect method. Hydrol Earth Syst Sci 24(3):1171–1187

Penna D, van Meerveld HJ (2019) Spatial variability in the isotopic composition of water in small catchments and its effect on hydrograph separation. Wiley Interdiscip Rev Water 6(5):e1367

Petermann E, Gibson JJ, Knöller K, Pannier T, Weiß H, Schubert M (2018) Determination of groundwater discharge rates and water residence time of groundwater-fed lakes by stable isotopes of water (18O, 2H) and radon (^{222}Rn) mass balances. Hydrol Process 32(6):805–816

Pozdniakov SP, Wang P, Grinevsky SO, Frolova NL (2022) A Physically based model of a two-pass digital filter for separating groundwater runoff from streamflow time series. Water Resour Res 58(3):e2021WR031333

Raymond PA, Zappa CJ, Butman D, Bott TL, Potter J, Mulholland P, Laursen AE, McDowell WH, Newbold D (2012) Scaling the gas transfer velocity and hydraulic geometry in streams and small rivers. Limnol Oceanogr Fluids Environ 2(1):41–53

Santos IR, de Weys J, Tait DR, Eyre BD (2013) The contribution of groundwater discharge to nutrient exports from a coastal catchment: post-flood seepage increases estuarine N/P ratios. Estuaries Coasts 36(1):56–73

Schallenberg M, de Winton MD, Verburg P, Kelly DJ, Hamill KD, Hamilton DP (2013) Ecosystem services of lakes. In: Ecosystem services in New Zealand: conditions and trends. Manaaki Whenua Press, Lincoln, pp 203–225

Schmidt S, Geyer T, Guttman J, Marei A, Ries F, Sauter M (2014) Characterisation and modelling of conduit restricted karst aquifers–example of the Auja spring, Jordan Valley. J Hydrol 511:750–763

Schubert M, Paschke A, Bednorz D, Bürkin W, Stieglitz T (2012) Kinetics of the water/air phase transition of radon and its implication on detection of radon-in-water concentrations: practical assessment of different on-site radon extraction methods. Environ Sci Technol 46(16):8945–8951

Schubert M, Knoeller K, Mueller C, Gilfedder B (2020a) Investigating river water/groundwater interaction along a rivulet section by ^{222}Rn mass balancing. Water 12(11):3027

Schubert M, Siebert C, Knoeller K, Roediger T, Schmidt A, Gilfedder B (2020b) Investigating groundwater discharge into a major river under low flow conditions based on a radon mass balance supported by tritium data. Water 12(10):2838

Shao G, Zhang D, Guan Y, Sadat MA, Huang F (2020) Application of different separation methods to investigate the baseflow characteristics of a Semi-Arid Sandy Area, Northwestern China. Water 12(2):434

Solgi A, Zarei H, Marofi S (2022) A new approach to use of wavelet transform for baseflow separation of Karst springs (case study: Gamasiyab spring). Appl Water Sci 12(12):1–15

Sterner RW, Keeler B, Polasky S, Poudel R, Rhude K, Rogers M (2020) Ecosystem services of Earth's largest freshwater lakes. Ecosyst Serv 41:101046

Sukanya S, Noble J, Joseph S (2021) Factors controlling the distribution of radon (^{222}Rn) in groundwater of a tropical mountainous river basin in southwest India. Chemosphere 263:128096

Sukanya S, Noble J, Joseph S (2022) Application of radon (^{222}Rn) as an environmental tracer in hydrogeological and geological investigations: an overview. Chemosphere 135141

Sun X, Du Y, Deng Y, Fan H, Tao Y, Ma T (2021) Contribution of groundwater discharge and associated contaminants input to Dongting Lake, Central China, using multiple tracers (^{222}Rn, 18O, Cl−). Environ Geochem Health 43(3):1239–1255

Tongal H, Booij MJ (2022) Simulated annealing coupled with a Naive Bayes model and base flow separation for streamflow simulation in a snow dominated basin. Stoch Environ Res Risk Assess 1–24

Unland NP, Cartwright I, Andersen MS, Rau GC, Reed J, Gilfedder BS, Atkinson AP, Hofmann H (2013) Investigating the spatio-temporal variability in groundwater and surface water interactions: a multi-technical approach. Hydrol Earth Syst Sci Discuss 10(3)

Wallace H, Wexler EJ, Malott S, Robinson CE (2021) Evaluating lacustrine groundwater discharge to a large glacial lake using regional scale radon-^{222} surveys and groundwater modelling. Hydrol Process 35(4):e14165

Wan C, Li K, Shen S, Gibson JJ, Ji K, Yi P, Yu Z (2019) Using tritium and 222 Rn to estimate groundwater discharge and thawing permafrost contributing to surface water in permafrost regions on Qinghai-Tibet Plateau. J Radioanal Nucl Chem 322:561–578

Wang Y, Zhang L, Wang J, Guo Q (2020) Study on an on-site radon-in-water measurement system based on degassing membrane. Radiat Meas 131:106231

Schroeder ED (2003) Water resources. Encyclopedia of physical science and technology, 3rd edn, pp 721–751

Webb JR, Santos IR, Tait DR, Sippo JZ, Macdonald BC, Robson B, Maher DT (2016) Divergent drivers of carbon dioxide and methane dynamics in an agricultural coastal floodplain: Post-flood hydrological and biological drivers. Chem Geol 440:313–325

Williams RT, Fryirs KA, Hose GC (2020) The hydrological function of a large chain-of-ponds: a wetland system with intermittent surface flows. Aquat Sci 82(3):61

Wilson J, Rocha C (2016) A combined remote sensing and multi-tracer approach for localising and assessing groundwater-lake interactions. Int J Appl Earth Obs Geoinf 44:195–204

Wilson AM, Evans TB, Moore WS, Schutte CA, Joye SB (2015) What time scales are important for monitoring tidally influenced submarine groundwater discharge? insights from a salt marsh. Water Resour Res 51(6):4198–4207

Wohl E (2021) An integrative conceptualization of floodplain storage. Rev Geophys 59(2):p.e2020RG000724

Xia Y, Li H, Wang B, Ma Z, Guo X, Zhao K, Zhao C (2022) Characterization of shallow groundwater circulation based on chemical kinetics: a case study of Xiong'an new area, China. Water 14(12):1880

Chapter 6
Radon in Submarine Groundwater Discharge Studies

6.1 Introduction

Submarine groundwater discharge (SGD) refers to the influx of fresh or saline groundwater into oceans and it may have a significant influence on biogeochemical balance and water budget of oceans (Moosdorf et al. 2015; Knee et al. 2016; Santos et al. 2021). Hence, the estimation of this freshwater component loaded with solutes (e.g., carbon, iron, silica, strontium, etc.) and nutrients is crucial (Luo et al. 2014; Chevis et al. 2015; Rahman et al. 2019; Damodararao and Singh 2022). Additionally, groundwater alkalinity buffers ocean acidification as does fresh SGD. Based on the nature of flux, Luijendijk et al. (2020) classified SGD into three—(i) fresh SGD, (ii) near-shore terrestrial groundwater/NGD, and (iii) recirculated sea water (Fig. 6.1).

Fresh SGD comprises meteoric groundwater end member discharge below mean sea level. When the meteoric groundwater discharge occurs above the mean sea level near the coastline, it is regarded as NGD. Fresh SGD and NGD depend on recharge from onshore precipitation whereas recirculated sea water is controlled by waves, tides, storm surges, and density-dependent flow. Processes which have been variously known as the hydrodynamic exchange (Wilson et al. 2016), benthic exchange (Duque et al. 2019), or porewater exchange (Lopez et al. 2020) also result in the exchange of groundwater and seawater. The impacts of currents and tides frequently coincide with wave-driven SGD.

A lion's share of studies has estimated fresh SGD fluxes at local scale. Owing to its highly dynamic nature, researchers often face uncertainty and bias in estimating a global budget of fresh SGD using local measurements. However, recent studies suggested that the global nutrient contribution of this component to marine systems is around 10% of river discharge and vary by approximately four orders of magnitude (Taniguchi et al. 2019; Luijendijk et al. 2020). Understanding near-shore terrestrial groundwater dynamics along with surface water budget calculations is necessary for effective monitoring of ecosystems. Luijendijk et al. (2020) applied numerical models of density-dependent groundwater flow to quantify the partitioning of terrestrial as

S. S. and S. Joseph, *Environmental Radon*, Environmental Science and Engineering,
https://doi.org/10.1007/978-981-99-2672-5_6

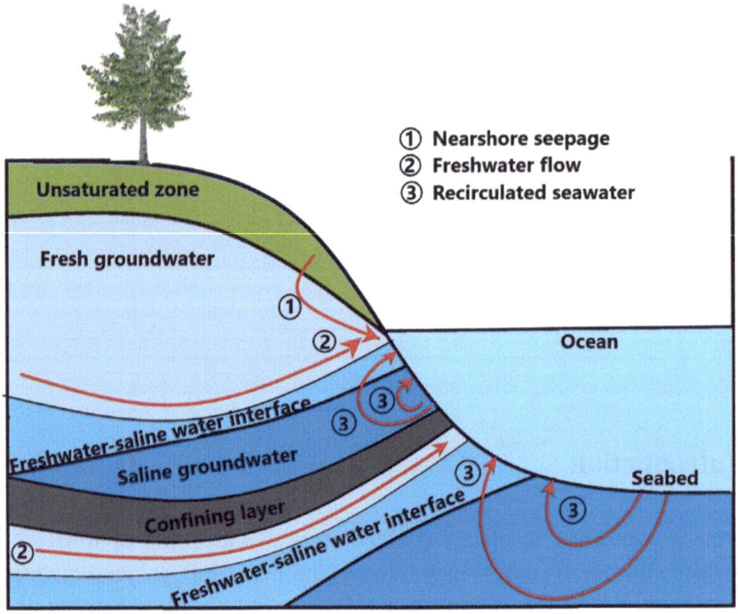

Fig. 6.1 Conceptual diagram of SGD pathways (*after* Swarzenski et al. 2004)

well as submarine groundwater discharge and the sensitivity of coastal groundwater discharge with respect to factors, viz., topography, permeability, recharge rate, and size of contributing area.

6.2 Significance of SGD Studies

For several years, the scientific community considered rivers, storm water, wastewater runoff, agricultural runoff, waste storage runoff, atmospheric deposition, etc. as the contributors of dissolved nutrients to the sea (Fredston-Hermann et al. 2016). However, later studies identified the terrestrial groundwater (flowing down-gradient, discharging into sea) as an important contributor capable of altering the chemical budget of ocean (Moosdorf and Oehler 2017; Cho et al. 2018; Mayfield et al. 2021).

For a coon's age, a misconception existed that these "invisible" pathways are insignificant in land-sea exchange. Due to this negligence, the science of SGD emerged not long ago. Over the recent years, research has been focusing on its chemical as well as volumetric (quantification) aspects. Isotope-based (^{223}Ra, ^{224}Ra, ^{226}Ra) cross-shore gradients in a Mediterranean Bay (Balearic Islands, Spain) allowed the estimation of SGD flow of 56,000 ± 13,000 m^3 d^{-1} (Rodellas et al. 2014). SGD could contribute disproportionately high quantities of dissolved inorganic and organic nutrients (Santos et al. 2021), although previous investigations have focused on rivers, streams, and oceanic drivers to evaluate nutrient budgets in coastal ecosystems. The biogeochemistry and functionality of coastal ecosystems could be adversely affected by the elevated nutrient contents in groundwater compared to surface water.

More often, the SGD-driven nutrient fluxes greater than or comparable to that of other natural (rivers, surface runoff, etc.) or anthropogenic inputs are observed along various coasts worldwide (Montiel et al. 2019; Correa et al. 2020; Douglas et al. 2021; Zhang et al. 2022). According to a recent study, the nutrients contributed via SGD exceeded to a considerably greater extent (>60%) compared to riverine nutrient inputs, with very high nitrogen-to-phosphorus ratio exceeding the Redfield ratio by around 76% (Santos et al. 2021). This ratio has been observed even in the case of continental margins and islands (Kwon et al. 2014; Moosdorf et al. 2015). Thus SGD is an integral component of coastal hydrological and biogeochemical cycles.

In addition to the estimation of SGD-driven nutrient fluxes, understanding the biogeochemical dynamics requires rigorous investigations on sources of groundwater nutrients and associated processes in the coastal ecosystems (Szymczycha and Pempkowiak 2015). The SGD–delivered nutrient load estimates into the Bay of Puck, southern Baltic Sea were found to be 49.9 ± 18.0 t yr^{-1} for dissolved inorganic nitrogen (DIN) and 56.3 ± 5.5 t yr^{-1} or PO_4- (Szymczycha et al. 2012). Hydrochemical analysis of groundwater significant discharge of nutrient loads, i.e., 4.27×10^6 mol day^{-1} for DIN, 2.24×10^4 mol day^{-1} for DIP, and 1.82×10^6 mol day^{-1} for DSi via SGD pathway in the shoreline of Xiangshan Bay, East China Sea (Wu et al. 2013).

The biological significance of SGD has been investigated in terms of primary productivity and eutrophication of estuarine and coastal ecosystems. In inadequately replenished water bodies, the input of these nutrients via SGD may be considerable enough to alter the dynamics and composition of microbial communities, induce algal blooms, and enhance hypoxic conditions (Hwang et al. 2005; Hu et al. 2006; Rodellas et al. 2017). Rodellas et al. (2014) highlighted the significance of SGD-derived nutrient and trace metal inputs in regulating the near-shore phytoplankton communities thriving in oligotrophic systems.

Despite being the essential micronutrients for planktonic community, heavy metals, viz., Fe, Mn, Zn, etc. in higher concentrations have toxic properties. Toxic heavy metals, viz., Cr, Cd, Hg, etc. could have considerable negative impacts on marine species diversity and ecosystems. Henceforth, quantifying the SGD-derived heavy metal fluxes is critical for marine environmental protection programs. In Bohai Bay (Bohai Sea, China), SGD-derived fluxes of heavy metals were found to be $(0.2–6.0) \times 10^7$ mol d^{-1} for Fe, $(1.2–2.7) \times 10^7$ mol d^{-1} for Mn, $(3.0–8.2) \times 10^5$ mol d^{-1} for Zn, $(2.7–7.4) \times 10^4$ mol d^{-1} for Cr, and $(0.6–1.8) \times 10^3$ mol d^{-1} for Cd, respectively (Wang et al. 2019a, b).

Recent advancements in this domain witnessed the investigation of SGD as a significant pathway for emerging contaminants other than heavy metals (Szymczycha et al. 2020; McKenzie et al. 2021a, b). Different contaminants impact ecosystems in different ways. For instance, PCPs like UV filters are capable of inducing coral bleaching and increase mortality rate of marine invertebrates (Araújo et al. 2018; He et al. 2019). They also interrupt with the normal reproductive functioning of marine invertebrates from sub-cellular to organism levels (Cuccaro et al. 2022). Parabens (e.g., butylparaben, ethylparaben, methylparaben, etc.) exposure activates several nuclear receptors altering hormone-dependent signaling pathways (Kang et al. 2019).

McKenzie et al. (2020) probed SGD-derived CECs, i.e., contaminants of emerging concern in Sydney Harbour (Australia) and the SGD rates suggested >80% of CEC inventories for one or multiple compounds, viz., caffeine, carbamazepine, dioxins, sulfamethoxazole, fluoroquinolones, and ibuprofen in almost all embayments. The study also confirmed the effect of mixing as a vital process driving the coastal inventories of persistent chemicals. In another similar contamination study, around 82% of target endocrine disrupting chemicals (EDCs) were found to be discharged into Laizhou Bay through SGD vector (Lu et al. 2020).

6.3 Factors Controlling SGD and Associated Pathways

Understanding the factors controlling SGD is vital for interpreting the dynamics of groundwater and its impact on the coastal ecosystem. Multiple factors and mechanisms driving SGD have been identified in the literature. The dynamic nature of SGD is highly influenced by groundwater-level fluctuations in coastal aquifer (hydraulic gradient), tidal cycles, and waves in ocean which consequently induce hydrological changes in SGD fluxes (Debnath and Mukherjee 2016; Lee et al. 2017; Jiang et al. 2021).

Density gradients, i.e., salinity gradients at the freshwater–saltwater boundary also control SGD (Michael et al. 2016) by means of buoyancy force. The upper saline plume, i.e., upper saline recirculation cell, overlies fresh groundwater which discharges near the low tide mark through a freshwater discharge tube. This upper saline plume forms an inverted salinity gradient. This density difference of freshwater and saline water parcels leads to diffusive instability and "salt fingering", a mixing process occurring when warm saline water overlies cold freshwater (Fang et al. 2021).

Greskowiak (2014) opined that increased attention must be taken while interpreting the estimates of field-based water and solute fluxes derived by SGD under salt fingering flow-shallow beach conditions. Similarly, thermal gradients also influence SGD (Bhagat et al. 2021; Young and Pradhanang 2021). Other factors, viz., ripple migration, sediment compaction, gas bubble upwelling, fluid shear, bioirrigation, and bioturbation and their possible impacts were noted by Santos et al. (2012).

Prolonged residence times of seawater component in sediments will have a stronger impact on the geochemical constitution of the exchanging seawater. Hence, it is crucial to delineate the relative contributions of the various physical processes that underlie saline SGD (Tamborski et al. 2017). While using geochemical techniques or field observations, often challenges are faced in estimating the different time scales of marine driving forces. Consequently, more than a few numerical models have been predominantly used to investigate the functions of currents, tides, waves, and density (Xin et al. 2015; Li et al. 2016; Luijendijk et al. 2020). However, numerical models incorporated with simplified geological structure may significantly underestimate saltwater circulation by calibrating salinity distributions or measured data of SGD which results in considerable uncertainties (Yu et al. 2022).

Since the discovery of the beach water table over-height or super-elevation and the subsequent finding of fresh groundwater tubes below upper saline plumes in beach aquifers, tidally driven SGD has been thoroughly explored by many researchers. At high tide, seawater seeps into marshes, cracked aquifers, and/or beaches, forming a circulation cell that propels the return of seawater to the ocean.

Based on the nature of water exchange processes, Garcia-Orellana et al. (2021) classified the SGD pathways into five different categories:

(i) **Terrestrial groundwater discharge**: It comprises mostly fresh groundwater and this pathway is controlled by the hydraulic gradient between land and sea. Discerning the various hydrological source contributions to oceanic environment via terrestrial groundwater discharge is vital in many aspects. For example, this conduit acts as the nutrient substrate for littoral to marine organisms, especially in the sub-littoral zone (Beck et al. 2017).

(ii) **Density-driven seawater circulation**: Density-driven or convective pathway is based on either density gradient along the freshwater–saltwater interface or thermo-haline gradients in permeable sediments.

This mixing zone (freshwater–saltwater boundary) creates a hotspot for reactions of the two distinct water bodies with different chemical properties (Robinson et al. 2018), and this density gradient determines the fate of both land-sourced as well as sea-borne chemicals (Pu et al. 2020).

(iii) **Seasonal exchange of seawater**: This flow path is driven by the movement of freshwater–saline water interface due to temporal variations in aquifer recharge or sea-level fluctuations. Precipitation and groundwater table may have impact on SGD flux with more or less uncertainties associated with meteorological parameters, viz., temperature, humidity, etc. (Zhang et al. 2017).

(iv) **Shoreface circulation of seawater**: It includes intertidal circulation driven by tidal inundation occurring at coastal environments including beach faces, salt marshes, mangroves etc. as well as wave-setup. A Fast Fourier Transfer (FFT) analysis in Osaka Bay (Japan) showed that the frequencies for SGD magnitude change corresponded to the dominant periods of 341.3 h (semi-monthly), 24.1 h (diurnal), and 12.3 h (semi-diurnal) for the sea-level oscillations induced by tide (Taniguchi et al. 2002). Tidal pumping was found to be the controlling mechanism responsible for magnitude variation of SGD through ebb–flood to neap–spring oscillation (Taniguchi et al. 2002; Hsu et al. 2020).

(v) **Porewater exchange or benthic flux**: This short-scale (cm scale) advective pathway is driven by distinct mechanisms such as current bedform interactions, bioirrigation, tidal and wave pumping, shear flow, and ripples migration (Guimond and Tamborski 2021; Murgulet et al. 2022).

While field experiments have not fully isolated wave-driven recirculation's relative contribution to total SGD, waves are known to play a significant role in fluid exchange. Numerical models are typically used in wave-driven SGD investigations in ideal, phase-averaged settings.

Waves substantially increase SGD by expanding the tidally driven upper saline plume in beaches. Waves can play a significant role in porewater exchange, with combined volumetric exchange rates reportedly outpacing tidal pumping by an order of magnitude.

Seawater infiltrating into beach sediments, fractured aquifers, or marshes during high tide eventually returns to the ocean during low tide within a temporal scale of few days to months (Santos et al. 2021). Density-driven seawater circulation is generally subjected to alteration due to upper saline plume (upper part of aquifer where seawater recirculation cell is observed) in permeable coastal aquifers (Robinson et al. 2018). Fang et al. (2022) observed that the upper saline plume is neither stable nor unstable in the intertidal zone. Instead, it is a dynamically stable–unstable process with respect to the seasonal variation of terrestrial freshwater input.

Most of the earlier SGD studies overlooked the effect of episodic events on water and nutrient fluxes. In order to narrow such gaps, Diego-Feliu et al. (2022) probed the relationship between an extreme precipitation event and SGD magnitude. Extreme precipitation event was found to be having pronounced effects on groundwater flow, i.e., 1 order of magnitude higher compared to those in baseflow conditions. Taking account of extreme episodic events in estimating SGD fluxes is relevant in the present scenario of climate change.

Of late, the influence of atmospheric groundwater forcing on SGD during storms and tropical cyclones has been receiving attention in the scientific community. The meteoric recharge as a result of winter storm/tropical cyclone-induced heavy precipitation could intensify SGD fluxes during those periods (Beebe et al. 2022). In another recent study, storm-induced precipitation events were found to be elevating subsurface *E. coli* microbial exports to coastal water via wave-induced recirculated seawater (Cheng et al. 2023).

6.4 Measurement Techniques

Earlier, one of the primary challenges associated with submarine groundwater discharge estimation was the difficulty in flow measurement. Unlike large rivers, gauging options were not available for measuring SGD fluxes. However, over the past few decades, there have been continuous efforts from researchers in developing techniques for estimating these flows. A short account on some of the conventional non-isotopic (e.g., seepage meters, geophysical) and isotopic techniques (e.g., radium and radon) are provided in the following sub-sections.

6.4.1 Conventional (Non-isotope) Techniques

Seepage meters are one of the traditionally applied tools in SGD studies. The primitive Lee-type manual seepage meters (Lee 1977) consist of a 55-gallon steel drum fitted with a sample port and a plastic collection bag. This drum with an open chamber at its end is inserted into bed sediments. When water seeps into this chamber through sediments, it is forced into the attached plastic collection bag. The SGD flux rate, i.e., seepage rate is estimated by the change in volume over a measured time interval. Advanced automated variants of this technique were developed subsequently in the later years. Few examples are heat pulse, continuous heat pulse, ultrasonic, electromagnetic seepage meters, etc. (Rosenberry et al. 2020; Duque et al. 2020).

One major disadvantage of seepage meters is that they require comparatively calm environment and strong breaking waves tend to dislodge these instruments. Lee et al. (2018) proposed a new design-buoy-type seepage meter which works on the principle of hydraulic head difference between seawater and groundwater at freshwater–saline water interface. A vertical cylinder is fixed to a buoy for deriving groundwater seepage (volumetric flow rate) corresponding to the difference of hydraulic head. While buoy installation seems to be effective under stronger wind and wave/current conditions, this system requires some technical improvements regarding electric power supply and sampling.

Geophysical methods in SGD studies are mostly based on temperature and salinity gradients. Compared to seepage meter technique, geophysical methods are more proficient in discriminating freshwater and seawater components of SGD. These techniques are more suitable for mapping SGD zones. For example, sea surface temperature mapping by means of thermal infrared (TIR) sensors identifies locations of groundwater discharge based on temperature anomalies or contrast between groundwater and seawater. Plumes of buoyant groundwater could be detected, especially in volcanic or karstic environments.

Geo-electric methods based on electrical conductivity or resistivity or tomography, etc. are applied using an array of multiple electrodes either inserted into sediment surface or towed behind a boat. Geo-electric mapping is useful in identifying ideal locations for installing seepage meters where preferential flow paths exist. Salinity-based geophysical methods could be used for beach-scale point measurement of SGD fluxes. This is a useful tool in establishing a subsurface salt balance model for estimating SGD fluxes.

High-resolution hydroacoustic seafloor mapping techniques using echosounders and sonar identify sites of seafloor structures (e.g., depressions) are associated with SGD. Echosounders (single beam or multibeam) gather bathymetric and backscatter data (Candio et al. 2022). Gas bubbles in seawater are potential acoustic scatterers. Hoffmann et al. (2020) used high-resolution acoustic data to characterize SGD-associated intrapockmarks (<15 m across) and the results suggested that strong multibeam backscatter signal indicates free shallow gas and advective groundwater flow in organic-rich muddy sediments of Eckernförde Bay (Germany).

Gabuyo and Siringan (2022) recommended low-cost, recreational-grade echosounders (RGES) as an alternative mapping technique for preliminary nearshore SGD research in bubble-filled discharge zones with shallow hydrothermal vents and methane emanations. Seismic profiling is another method for delineating the geological evolution of SGD and its flow paths. Additionally, seismic profiling data reveals information on submarine paleochannels and hydrological connectivity of coastal aquifers with offshore discharge sites. Seismic reflection data often helps researchers to overcome the challenge of extrapolating subsurface properties where data availability is sparse for geological modeling (Thomas et al. 2022).

6.4.2 *Isotope Techniques*

Isotope tracers are the most commonly used tools in geochemical SGD estimation. Small-scale spatial and temporal variability is a limitation in the case regarding most of the conventional techniques. These variations get smoothed out over time and space by using isotope tracers. Radium (^{223}Ra, ^{224}Ra, ^{226}Ra, ^{228}Ra) and radon (^{222}Rn) radioactive isotopes are the most frequently applied tracers in SGD studies. However, the sources, sinks, residence times of water component, boundary conditions, etc. should be clearly understood while using these techniques to avoid overestimation or underestimation of SGD estimates. Another major challenge in quantifying global-scale SGD has been the lack of reliable techniques for determining the flux at specific sites which could be overcome by the use of isotopic tracers.

As groundwater is the dominant source of radioactive isotopes, this clear distinct signature could be used to differentiate seawater and other components like river water, rainwater, etc. For instance, the different half-lives of radium isotopes could be used in separating SGD from porewater exchange (Tamborski et al. 2017). Most of the wave-driven flow processes are active in the upper part (~2–12 cm) of sandy seafloor sediments (Jahnke 2005). Under such conditions, hydrodynamic exchange is unlikely to explain the observed radium enrichments (^{228}Ra/^{226}Ra activity ratios), as the levels of ^{226}Ra and ^{228}Ra generated in sandy sediments are negligible (George et al. 2020; Sanial et al. 2021).

The forthcoming section explains the suitability of radon and methodology for estimating SGD fluxes.

6.5 Radon as SGD Tracer

The presence of radon (^{222}Rn) in surface water is extremely low, but it is produced in the aquifer matrix as a result of the radioactive decay of ^{226}Ra. Hence, the activity concentration of radon is typically three to five times higher in groundwater compared to surface water (sea water in this context). Hence, the study of radon distribution patterns is relevant to the investigation of localized SGD in the coastal system. It is possible to identify ^{222}Rn at extremely low activity concentrations, even after extensive diluting in coastal waters.

Overall, radon's conservative character, short half-life, high abundance in groundwater relative to surface water, and the possibility for in situ measurement make it a great tracer for locating areas of large groundwater discharge in coastal water. Water and associated matter or nutrient flows can be measured with the help of radon mass balances within defined SGD impacted areas of shoreline.

A problem with ^{222}Rn is that it is a gas that easily escapes into the atmosphere due to its low vapor pressure. SGD can be effectively quantified through continuous monitoring of ^{222}Rn in coastal waters (Burnett and Dulaiova 2003). A conceptual model (Fig. 6.2) for estimating SGD using the mass balance of ^{222}Rn in the coastal zone was developed by Burnett and Dulaiova (2003). This method works on the premise that changes in ^{222}Rn activities can be converted to fluxes through long-term monitoring of ^{222}Rn concentrations in coastal water, with advective transport of groundwater through the sediment assumed to be the dominant process for the observed ^{222}Rn, and losses due to atmospheric evasion and mixing with offshore waters and gains due to diffusion from the bottom sediments accounted for.

To assess SGD, we can use the following steady-state ^{222}Rn mass balance equation for coastal water:

$$F_{SGD} = F_{atm} + F_{decay} + F_{mix} - F_{sed} \tag{6.1}$$

where F_{SGD} is the ^{222}Rn flux attributed to SGD, F_{atm} is the ^{222}Rn flux lost to the atmosphere, F_{decay} is the ^{222}Rn flux lost to radioactive decay within the system, F_{mix} is the ^{222}Rn flux lost to the open ocean through mixing, and F_{sed} is the ^{222}Rn flux diffused into the coastal water. The fluxes of ^{222}Rn are all given in Bqm-2d-1 (i.e., per unit area).

Significant radon losses from surface water bodies are typically caused by radon degassing (F_{atm}) over the air–water interface. Its velocity is determined by the concentration gradient at the water/air contact, which drives molecular diffusion and the strength of the turbulent gas transfer.

It can be estimated as

$$F_{atm} = k(C_w - \alpha(T)C_a) \tag{6.2}$$

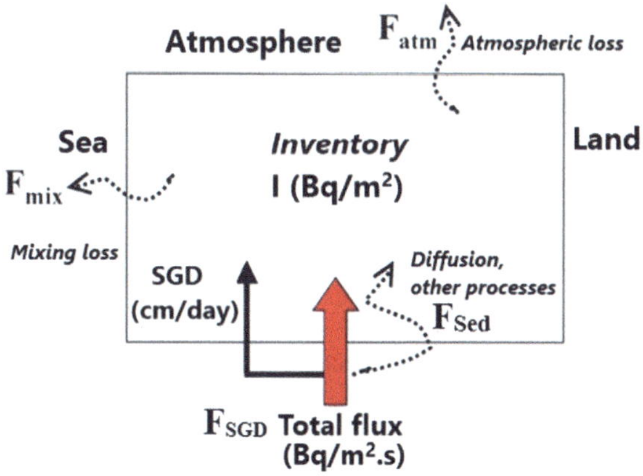

Fig. 6.2 Conceptual model for estimating submarine groundwater discharge in the coastal zone from the continuous measurement of ^{222}Rn in coastal water (*Source* Sukanya et al. 2022)

where k is the gas transfer coefficient (m/d); C_w and C_a are the ^{222}Rn activity concentrations in water and air, respectively (Bq/m^3); $\alpha(T)$ is the partition coefficient of radon between water and air (dimensionless), which depends upon the temperature and is estimated using Eq. 6.2.

The gas exchange velocity is usually estimated using an empirical equation:

$$k = 0.45\mu_{10}^{1.6} \times \left(\frac{S_c}{600}\right)^{-b} \tag{6.3}$$

where μ_{10} is the wind speed at 10 m above the water surface; S_c is the Schmidt number (dimensionless), the Schmidt number is equal to the radon molecular diffusion coefficient in water divided by the kinematic viscosity of water at 20 °C and 101.325 kPa (1.0043 × 10^{-6} m^2/s and 1.16 × 10^{-9} m^2, respectively). The Schmidt number of carbon dioxide at 20 °C is 600, hence most comparisons are made using a standardization factor of b = 0.5 when the wind speed is greater than 3.6 m/s and b = 0.667 when the wind speed is less than 3.6 m/s.

μ_{10} can be expressed as

$$\mu_{10} = \mu_h(0.097 \ \ln \left(\frac{h}{10}\right) + 1)^{-1} \tag{6.4}$$

where μ_h is the wind speed at the measurement height (m/s) and h is the wind observation elevation above the water surface (m).

$$S_c = 3417.6 \times e^{-0.0634} \times T_a \tag{6.5}$$

The loss of ^{222}Rn flux from the coastal water due to radioactive decay (F_{decay}) is given by

$$F_{decay} = I_{cw} \times \lambda \tag{6.6}$$

where I_{cw} is the ^{222}Rn inventory in the coastal water (Bq/m^2) and λ is the decay constant of ^{222}Rn (0.186 d^{-1}).

Burnett and Dulaiova (2003) proposed an equation using excess ^{222}Rn inventory to estimate the tidal effect on ^{222}Rn activity concentration, where excess ^{222}Rn inventory is defined as the product of excess ^{222}Rn in water (^{222}Rn–^{226}Ra activity concentration) and the water depth (d). The equation is as follows:

$$I_{cw} = (C_w - C^{226}Ra) \times d \tag{6.7}$$

where C^{226}Ra is the ^{226}Ra activity concentration (Bq/m^3) in the coastal water.

Many studies have concluded that ^{226}Ra is more tightly bonded to the sediments than to water, therefore the radioactive decay of the parent isotope within this system, which would produce radon, is insignificant.

The loss of ^{222}Rn flux from the coastal water via exchange with the open ocean (F_{mix}) is given by

$$F_{mix} = C_w \times \lambda_{mix} \times h \quad (6.8)$$

where C_w is the ^{222}Rn activity concentration in the system and λ_{mix} is the exchange rate between the coastal water and the open sea (d^{-1}).

The ^{222}Rn flux entering into the coastal water (per unit area) due to the diffusion from the sediments (F_{sed}) is given by (Sun et al. 2021)

$$F_{sed} = (\lambda \times D_{sed})^{0.5}\left(C_p - C_w\right) \quad (6.9)$$

where D_{sed} (m^2/d) is the molecular diffusion coefficient of ^{222}Rn in wet bulk sediment and C_p (Bq/m^3) is the average ^{222}Rn activity concentration in the pore water of the sediments.

The D_{sed} depends on the molecular diffusion coefficient (D_m) in coastal water and the porosity (n) of the sediment, and D_m is dependent on temperature, and it is described by the following equation:

$$D_{sed} = \frac{D_m}{[1 - \ln(n^2)]^2} \quad (6.10)$$

$$-\log D_m = \left(\frac{980}{T}\right) + 1.59 \quad (6.11)$$

Sediment equilibrium incubation experiments are commonly used to determine the ^{222}Rn activity concentration in sediment pore water by placing 150 g of sediment and 500 mL of in situ collected surface water in a cone-shaped bottle, sealing the bottle, and incubating the mixture for 30 days, until equilibrium of radon activity concentration is reached (Corbett et al. 1998).

From the incubation experiment, the ^{222}Rn activity concentration in the pore water of the sediment (C_p) can be calculated from the measured ^{222}Rn activity concentration of overlying water at equilibrium combined with physical parameters of sediment:

$$C_p = \frac{C_{sed} \times \rho_w}{n} \quad (6.12)$$

$$C_{sed} = \frac{A_0 \times V_0}{M_{sed}} \quad (6.13)$$

where C_{sed} is the ^{222}Rn content in unit volume of the wet sediment (Bq/kg), ρ_w is wet density of sediment (kg/m^3), n is the porosity of sediment, A_0 is the ^{222}Rn activity concentration in the overlying water (Bq/m^3) at equilibrium, V_0 is the volume (m^3) of the overlying seawater used, and M_{sed} is the sediment mass (kg) used in the incubation experiment.

Using the calculated F_{SGD} from Eq. 6.1, the SGD rate can be estimated as

$$\psi_{\text{SGD}} = \frac{F_{\text{SGD}}}{\left(C_g \times A_{\text{Bott}}\right)} \tag{6.14}$$

where ψ_{SGD} is the seepage rate of submarine groundwater (m/d), C_g is the average ^{222}Rn activity concentration in groundwater (Bq/m^3), and A_{Bott} is the bottom area of the bay (m^2).

Under the transient condition, the mass balance Eq. (6.1) can be modified as

$$F_{\text{SGD}} = \frac{I_{cw}}{\Delta_t} + F_{\text{atm}} + F_{\text{decay}} + F_{\text{mix}} - F_{\text{sed}} \tag{6.15}$$

where $\frac{I_{cw}}{\Delta_t}$ is the change in activity concentrations of excess ^{222}Rn between 2 successive hours of measurement.

While the radon mass balance approach is useful for estimating SGD fluxes, it is not applicable for fractionalizing SGD into freshwater (f_{FSGD}) and recirculated seawater discharge (f_{RSGD}) within the area of interest (AOI). The salt mass balance proposed by Schubert et al. (2019) assumes that FSGD is the only source of freshwater for AOI. Before and during the sample campaigns, it was believed that other FSGD sources such as precipitation, surface water runoff, and evaporation would have a minor impact. Another assumption was that RSGD does not affect the salinity of the AOI's water. The average salinity of the water within the AOI (Sal_{AOI}) and the salinity of the offshore seawater end member (Sal_{SW}) were used to calculate the FSGD fraction (f_{FSGD}) under these conditions:

$$f_{\text{FSGD}} = \frac{1 - \text{Sal}_{\text{AOI}}}{\text{Sal}_{\text{SW}}} \tag{6.16}$$

Following this, FSGD flux (E_{FSGD}) is required for sustaining f_{FSGD} assuming offshore mixing as the sole process removing freshwater from the AOI. Using mixing rate and water volume of AOI (V_{AOI}), FSGD flux was estimated based on FSGD fraction:

$$E_{\text{FSGD}} = \left(1 - e^{(-\text{max})}\right) \times V_{\text{AOI}} \times f_{\text{FSGD}} \tag{6.17}$$

In a study by Wu et al. (2013), continuous monitoring of ^{222}Rn was used to estimate submarine groundwater discharge (SGD) and associated nutrient fluxes into Xiangshan Bay, East China Sea. A stationary time-series mooring platform in the form of a floating wooden house was used to apply a ^{222}Rn balance model for the assessment of SGD. The study revealed a clear inverse relationship between tidal height and ^{222}Rn activity concentration. A ^{222}Rn mass balance model was employed to estimate the ^{222}Rn flux in the study area, which enabled quantification of groundwater discharges to the bay. The model accounted for loss and gain of ^{222}Rn due to tidal effects, atmospheric evasion, sediment diffusion, and mixing loss to the open sea.

Urquidi-Gaume et al. (2016) conducted an intensive time series of continuous radon measurements at two sites (Balandra and Merito) in the southeastern portion of La Paz Bay, an arid region where surface flow contributions to nutrient transport are negligible. The study aimed to quantify the fluxes of submarine groundwater discharge (SGD) and identify the SGD components responsible for nutrient transport. Strong radon signals were observed at both sites, and using radon signatures and mass balance, the authors found that the SGD components responsible for nutrient transport were different at the two sites. Fresh SGD was the dominant component in Balandra, whereas saline SGD controlled the Merito Region.

Wang et al. (2017) estimated SGD rates at two sites in Daya Bay, South China Sea, using a radon mass balance model. Daya Bay is a semi-closed bay that is heavily influenced by anthropogenic activities. The study aimed to investigate the influence of SGD on nutrient fluxes and eutrophication in the bay. The authors found that the nutrient fluxes influenced by SGD were considerable, with dissolved nitrates at a rate of 3.28×10^5 mol/d, dissolved inorganic phosphorus at 5.84×10^3 mol/d, and reactive silicate at 8.97×10^5 mol/d. The N:P ratio was found to be high (~43), indicating eutrophication.

Radon mass balance approach is a widely used technique to quantify submarine groundwater discharge (SGD) fluxes. However, one of its limitations is the uncertainty associated with the calculation of endmembers, which are mostly based on data from a limited number of samples in regional scale studies. This is particularly challenging in locations with significant spatial heterogeneity in radon distributions.

To address this issue, Schubert et al. (2019) developed a stochastic Monte Carlo simulation approach that employs probability density functions (pdf). While assuming uncertainties as Gaussian distributions, the input parameters used in the mass balance calculations were assumed to follow a lognormal distribution that replicates the natural system's distribution. This approach was used to reduce uncertainties in SGD flux estimation by generating a large number of random realizations of the input parameters, thus providing more realistic estimates of the endmembers and a range of possible SGD flux values.

6.6 Limitations and Uncertainties Associated with Radon Tracer

While ^{222}Rn serves as a potential tracer in tracking and quantifying submarine groundwater discharge fluxes, uncertainties may occur in shallow wind-exposed coastal settings exposed to shallow wind. Schubert et al. (2022) noted that the reliability of radon tracer might be questionable at such locations where wind-driven water circulation could facilitate offshore mixing, i.e., mixing of radon-rich coastal waters with low radon offshore waters. Hence, wind conditions should be taken into consideration prior to radon-based SGD studies.

In spite of the potential of radon as a tracer in SGD studies, various hydrogeological conditions such as the interface between seawater and freshwater, temporal variations, origin and residence time of groundwater, and the use of conceptual models in sampling design and frequency can introduce uncertainties. To obtain accurate estimates of SGD, it is necessary to conduct uncertainty analysis to mitigate these uncertainties (Duque et al. 2019). Conceptual uncertainties associated with the radon mass balance model arise from the inherent assumptions regarding the prevailing conditions and the choice of the equation for flux estimation (Rodellas et al. 2021; Coluccio et al. 2021).

The spatial distribution of tracer using Langrangian sampling and water age profile is a critical factor in correcting the effects of physical processes, viz., atmospheric loss, decay, etc. that determine the distribution of ^{222}Rn (Lopez et al. 2020; Zhang et al. 2020; Murgulet et al. 2022; Hébert et al. 2022). This correction is particularly important in the case of possible uncertainties in SGD estimation resulting from spatial factors. For instance, sampling locations far away from the region of higher water retention, as underestimation of SGD and overestimation of flushing rates may occur (Savatier and Rocha 2021). By providing a consistent magnitude of SGD at any given space and time, this information can be useful in understanding the complex hydrological processes in marine environments.

6.7 Conclusion and Outlooks

Submarine groundwater discharge (SGD) plays a significant role in the biogeochemical cycling and water budget of oceans. Quantification of SGD fluxes is often challenging due to its non-linear complex pathways as well as its highly variable nature with respect to space and time. Further complications occur when the system is influenced by rapidly changing wind patterns and water column. Continuous automated high-resolution time-series radon measurements may be carried out for reducing uncertainties when estimating SGD.

Improved versions of established radon detectors, viz., RAD7 should be developed. For instance, requirement of drying units is often one of the setbacks in continuous sampling surveys. This could be rectified by developing low-cost, energy-efficient instruments (with more sustainable batteries), which do not depend on drying units for reducing uncertainties associated with humidity. As the disequilibrium between ^{222}Rn and ^{226}Ra pair forms the principle for conducting SGD studies, an integrated instrument with simultaneous measurement facilities for both ^{222}Rn and ^{226}Ra isotopes should be designed.

References

Araújo MJ, Rocha RJM, Soares AMVM, Benedé JL, Chisvert A, Monteiro MS (2018) Effects of UV filter 4-methylbenzylidene camphor during early development of Solea senegalensis Kaup, 1858. Sci Total Environ 628:1395–1404

Beck AJ, Chambers RM, Mitchell MM, Bilkovic DM (2017) Evaluation of living shoreline marshes as a tool for reducing nitrogen pollution in coastal systems. In: Living shorelines. CRC Press, pp 271–290

Beebe DA, Huettemann MB, Webb BM, Jackson Jr WT (2022) Atmospheric groundwater forcing of a subterranean estuary: a seasonal seawater recirculation process. Geophys Res Lett 49(7):e2021GL096154

Bhagat C, Khandekar A, Singh A, Mohapatra PK, Kumar M (2021) Delineation of submarine groundwater discharge and seawater intrusion zones using anomalies in the field water quality parameters, groundwater level fluctuation and sea surface temperature along the Gujarat coast of India. J Environ Manag 296:113176

Burnett WC, Dulaiova H (2003) Estimating the dynamics of groundwater input into the coastal zone via continuous radon-222 measurements. J Environ Radioact 69(1–2):21–35

Candio S, Hoy S, Morrow T, Wilkins C, Copeland A (2022) NOAA ship Okeanos explorer mapping systems readiness report 2022

Cheng KH, Luo X, Jiao JJ, Yu S (2023) Storm accelerated subsurface Escherichia coli growth and exports to coastal waters. J Hazard Mater 441:129893

Chevis DA, Johannesson KH, Burdige DJ, Tang J, Moran SB, Kelly RP (2015) Submarine groundwater discharge of rare earth elements to a tidally-mixed estuary in Southern Rhode Island. Chem Geol 397:128–142

Cho HM, Kim G, Kwon EY, Moosdorf N, Garcia-Orellana J, Santos IR (2018) Radium tracing nutrient inputs through submarine groundwater discharge in the global ocean. Sci Rep 8(1):1–7

Coluccio KM, Santos IR, Jeffrey LC, Morgan LK (2021) Groundwater discharge rates and uncertainties in a coastal lagoon using a radon mass balance. J Hydrol 598:126436

Corbett D, Burnett W, Cable P, Clark S (1998) A multiple approach to the determination of radon fluxes from sediments. J Radioanal Nucl Chem 236(1–2):247–253

Correa RE, Tait DR, Sanders CJ, Conrad SR, Harrison D, Tucker JP, Reading MJ, Santos IR (2020) Submarine groundwater discharge and associated nutrient and carbon inputs into Sydney Harbour (Australia). J Hydrol 580:124262

Cuccaro A, De Marchi L, Oliva M, Battaglia F, Meucci V, Fumagalli G, Freitas R, Pretti C (2022) Ecotoxicological effects of the UV-filter 4-MBC on sperms and adults of the mussel Mytilus galloprovincialis. Environ Res 213:113739

Damodararao K, Singh SK (2022) Substantial submarine groundwater discharge in the estuaries of the east coast of India and its impact on marine strontium budget. Geochim Cosmochim Acta 324:66–85

Debnath P, Mukherjee A (2016) Quantification of tidally-influenced seasonal groundwater discharge to the Bay of Bengal by seepage meter study. J Hydrol 537:106–116

Diego- M, Rodellas V, Alorda-Kleinglass A, Saaltink M, Folch A, Garcia-Orellana J (2022) Extreme precipitation events induce high fluxes of groundwater and associated nutrients to coastal ocean. Hydrol Earth Syst Sci 26(18):4619–4635

Douglas AR, Murgulet D, Montagna PA (2021) Hydroclimatic variability drives submarine groundwater discharge and nutrient fluxes in an anthropogenically disturbed, semi-arid estuary. Sci Total Environ 755:142574

Duque C, Knee KL, Russoniello CJ, Sherif M, Risha UAA, Sturchio NC, Michael HA (2019) Hydrogeological processes and near shore spatial variability of radium and radon isotopes for the characterization of submarine groundwater discharge. J Hydrol 579:124192

Duque C, Russoniello CJ, Rosenberry DO (2020) History and evolution of seepage meters for quantifying flow between groundwater and surface water: Part 2—marine settings and submarine groundwater discharge. Earth Sci Rev 204:103168

Fang Y, Zheng T, Zheng X, Yang H, Wang H, Walther M (2021) Influence of tide-induced unstable flow on seawater intrusion and submarine groundwater discharge. Water Resour Res 57(4):e2020WR029038

Fang Y, Zheng T, Wang H, Zheng X, Walther M (2022) Influence of dynamically stable-unstable flow on seawater intrusion and submarine groundwater discharge over tidal and seasonal cycles. J Geophys Res Oceans 127(4):e2021JC018209

Fredston A, Brown CJ, Albert S, Klein CJ, Mangubhai S, Nelson JL, Teneva L, Wenger A, Gaines SD, Halpern BS (2016) Where does river runoff matter for coastal marine conservation? Front Mar Sci 3:273

Gabuyo MRP, Siringan FP (2022) Utility of low-cost recreational-grade echosounders in imaging and characterizing bubbly coastal submarine groundwater discharge. J Hydrol X 14:100118

Garcia-Orellana J, Rodellas V, Tamborski J, Diego- M, van Beek P, Weinstein Y, Charette M, Alorda-Kleinglass A, Michael HA, Stieglitz T, Scholten J (2021) Radium isotopes as submarine groundwater discharge (SGD) tracers: review and recommendations. Earth Sci Rev 220:103681

George C, Moore WS, White SM, Smoak E, Joye SB, Leier A, Wilson AM (2020) A new mechanism for submarine groundwater discharge from continental shelves. Water Resour Res 56(11):p.e2019WR026866

Greskowiak J (2014) Tide-induced salt-fingering flow during submarine groundwater discharge. Geophys Res Lett 41(18):6413–6419

Guimond J, Tamborski J (2021) Salt marsh hydrogeology: a review. Water 13(4):543

He T, Tsui MMP, Tan CJ, Ma CY, Yiu SKF, Wang LH, Fan TY, Lam PKS, Murphy MB (2019) Toxicological effects of two organic ultraviolet filters and a related commercial sunscreen product in adult corals. Environ Pollut 245:462–471

Hébert AJ, Flamand A, Chaillou G (2022) Origins and transformations of terrigenous dissolved organic matter in a transgressive coastal system. Estuar Coast Shelf Sci 279:108137

Hoffmann JJL, Schneider von Deimling J, Schröder JF, Schmidt M, Held P, Crutchley GJ, Scholten J, Gorman AR (2020) Complex eyed pockmarks and submarine groundwater discharge revealed by acoustic data and sediment cores in Eckernförde Bay, SW Baltic Sea. Geochem Geophys Geosystems 21(4):p.e2019GC008825

Hsu FH, Su CC, Wang PL, Lin IT (2020) Temporal variations of submarine groundwater discharge into a tide-dominated coastal wetland (Gaomei wetland, Western Taiwan) indicated by radon and radium isotopes. Water 12(6):1806

Hu C, Muller-Karger FE, Swarzenski PW (2006) Hurricanes, submarine groundwater discharge, and Florida's red tides. Geophys Res Lett 33(11)

Huettel M, Ziebis W, Forster S (1996) Flow-induced uptake of particulate matter in permeable sediments. Limnol Oceanogr 41(2):309–322

Huettel M, Berg P, Kostka JE (2014) Benthic exchange and biogeochemical cycling in permeable sediments. Ann Rev Mar Sci 6(1):23–51

Hwang DW, Lee YW, Kim G (2005) Large submarine groundwater discharge and benthic eutrophication in Bangdu Bay on volcanic Jeju Island, Korea. Limnol Oceanogr 50(5):1393–1403

Jahnke RA (2005) Matter cycling in coastal sediments. Glob Coast Ocean Multiscale Interdiscip Processes 11:163

Jiang ZP, Lv J, Li Q, Dai M, Kao SJ, Zheng N, Fan W (2021) Tidal-driven submarine groundwater discharge and its influences on the carbonate system of a coastal coral reef in the Northern South China Sea. J Geophys Res Oceans 126(7):e2021JC017203

Kang HM, Kim MS, Hwang UK, Jeong CB, Lee JS (2019) Effects of methylparaben, ethylparaben, and propylparaben on life parameters and sex ratio in the marine copepod Tigriopus japonicus. Chemosphere 226:388–394

Knee KL, Crook ED, Hench JL, Leichter JJ, Paytan A (2016) Assessment of submarine groundwater discharge (SGD) as a source of dissolved radium and nutrients to Moorea (French Polynesia) coastal waters. Estuaries Coasts 39(6):1651–1668

Kwon EY, Kim G, Primeau F, Moore WS, Cho HM, DeVries T, Sarmiento JL, Charette MA, Cho YK (2014) Global estimate of submarine groundwater discharge based on an observationally constrained radium isotope model. Geophys Res Lett 41(23):8438–8444

Lee DR (1977) A device for measuring seepage flux in lakes and estuaries. Limnol Oceanogr 22(1):140–147

Lee E, Shin D, Hyun SP, Ko KS, Moon HS, Koh DC, Ha K, Kim BY (2017) Periodic change in coastal microbial community structure associated with submarine groundwater discharge and tidal fluctuation. Limnol Oceanogr 62(2):437–451

Lee BJ, Lee JH, Kim DH (2018) An alternative approach to conventional seepage meters: Buoy-type seepage meter. Limnol Oceanogr Methods 16(5):299–308

Li X, Hu BX, Tong J (2016) Numerical study on tide-driven submarine groundwater discharge and seawater recirculation in heterogeneous aquifers. Stoch Env Res Risk Assess 30(6):1741–1755

Lopez CV, Murgulet D, Santos IR (2020) Radioactive and stable isotope measurements reveal saline submarine groundwater discharge in a semiarid estuary. J Hydrol 590:125395

Lu J, Wu J, Zhang C, Zhang Y (2020) Possible effect of submarine groundwater discharge on the pollution of coastal water: occurrence, source, and risks of endocrine disrupting chemicals in coastal groundwater and adjacent seawater influenced by reclaimed water irrigation. Chemosphere 250:126323

Luijendijk E, Gleeson T, Moosdorf N (2020) Fresh groundwater discharge insignificant for the world's oceans but important for coastal ecosystems. Nat Commun 11(1):1–12

Luo X, Jiao JJ, Moore WS, Lee CM (2014) Submarine groundwater discharge estimation in an urbanized embayment in Hong Kong via short-lived radium isotopes and its implication of nutrient loadings and primary production. Mar Pollut Bull 82(1–2):144–154

Mayfield KK, Eisenhauer A, Santiago Ramos DP, Higgins JA, Horner TJ, Auro M, Magna T, Moosdorf N, Charette MA, Gonneea ME, Brady CE (2021) Groundwater discharge impacts marine isotope budgets of Li, Mg, Ca, Sr, and Ba. Nat Commun 12(1):1–9

McKenzie T, Holloway C, Dulai H, Tucker JP, Sugimoto R, Nakajima T, Harada K, Santos IR (2020) Submarine groundwater discharge: a previously undocumented source of contaminants of emerging concern to the coastal ocean (Sydney, Australia). Mar Pollut Bull 160:111519

McKenzie T, Dulai H, Fuleky P (2021a) Traditional and novel time-series approaches reveal submarine groundwater discharge dynamics under baseline and extreme event conditions. Sci Rep 11(1):1–14

McKenzie T, Habel S, Dulai H (2021b) Sea-level rise drives wastewater leakage to coastal waters and storm drains. Limnol Oceanogr Lett 6(3):154–163

Michael HA, Scott KC, Koneshloo M, Yu X, Khan MR, Li K (2016) Geologic influence on groundwater salinity drives large seawater circulation through the continental shelf. Geophys Res Lett 43(20):10–782

Montiel D, Lamore AF, Stewart J, Lambert WJ, Honeck J, Lu Y, Warren O, Adyasari D, Moosdorf N, Dimova N (2019) Natural groundwater nutrient fluxes exceed anthropogenic inputs in an ecologically impacted estuary: lessons learned from Mobile Bay, Alabama. Biogeochemistry. 145(1):1–33

Moosdorf N, Oehler T (2017) Societal use of fresh submarine groundwater discharge: an overlooked water resource. Earth Sci Rev 171:338–348

Moosdorf N, Stieglitz T, Waska H, Dürr HH, Hartmann J (2015) Submarine groundwater discharge from tropical islands: a review. Grundwasser 20(1):53–67

Murgulet D, Lopez CV, Douglas AR (2022) Radioactive and stable isotopes reveal variations in nearshore submarine groundwater discharge composition and magnitude across low inflow northwestern Gulf of Mexico estuaries. Sci Total Environ 823:153814

Pu L, Xin P, Nguyen TT, Yu X, Li L, Barry DA (2020) Thermal effects on flow and salinity distributions in coastal confined aquifers. Water Resour Res 56(10):e2020WR027582

Rahman S, Tamborski JJ, Charette MA, Cochran JK (2019) Dissolved silica in the subterranean estuary and the impact of submarine groundwater discharge on the global marine silica budget. Mar Chem 208:29–42

Robinson CE, Xin P, Santos IR, Charette MA, Li L, Barry DA (2018) Groundwater dynamics in subterranean estuaries of coastal unconfined aquifers: controls on submarine groundwater discharge and chemical inputs to the ocean. Adv Water Resour 115:315–331

Rodellas V, Garcia-Orellana J, Tovar-Sánchez A, Basterretxea G, López-Garcia JM, Sánchez- D, Garcia- E, Masqué P (2014) Submarine groundwater discharge as a source of nutrients and trace metals in a Mediterranean bay (Palma Beach, Balearic Islands). Mar Chem 160:56–66

Rodellas V, Garcia-Orellana J, Trezzi G, Masqué P, Stieglitz TC, Bokuniewicz H, Cochran JK, Berdalet E (2017) Using the radium quartet to quantify submarine groundwater discharge and porewater exchange. Geochim Cosmochim Acta 196:58–73

Rodellas V, Stieglitz TC, Tamborski JJ, van Beek P, Andrisoa A, Cook PG (2021) Conceptual uncertainties in groundwater and porewater fluxes estimated by radon and radium mass balances. Limnol Oceanogr 66(4):1237–1255

Rosenberry DO, Duque C, Lee DR (2020) History and evolution of seepage meters for quantifying flow between groundwater and surface water: part 1—freshwater settings. Earth Sci Rev 204:103167

Sanial V, Moore WS, Shiller AM (2021) Does a bottom-up mechanism promote hypoxia in the Mississippi Bight? Mar Chem 235:104007

Santos IR, Eyre BD, Huettel M (2012) The driving forces of porewater and groundwater flow in permeable coastal sediments: a review. Estuar Coast Shelf Sci 98:1–15

Santos IR, Chen X, Lecher AL, Sawyer AH, Moosdorf N, Rodellas V, Tamborski J, Cho HM, Dimova N, Sugimoto R, Bonaglia S (2021) Submarine groundwater discharge impacts on coastal nutrient biogeochemistry. Nat Rev Earth Environ 2(5):307–323

Savatier M, Rocha C (2021) Rethinking tracer-based (Ra, Rn, salinity) approaches to estimate point-source submarine groundwater discharge (SGD) into coastal systems. J Hydrol 598:126247

Schubert M, Petermann E, Stollberg R, Gebel M, Scholten J, Knöller K, Lorz C, Glück F, Riemann K, Weiß H (2019) Improved approach for the investigation of submarine groundwater discharge by means of radon mapping and radon mass balancing. Water 11:749

Schubert M, Scholten J, Kreuzburg M, Petermann E, de Paiva ML, Köhler D, Liebetrau V, Rapaglia J, Schlüter M (2022) Radon (^{222}Rn) as tracer for submarine groundwater discharge investigation—limitations of the approach at shallow wind-exposed coastal settings. Environ Monit Assess 194(10):1–16

Sukanya S, Noble J, Joseph S (2022) Application of radon (^{222}Rn) as an environmental tracer in hydrogeological and geological investigations: an overview. Chemosphere 303:135141

Sun X, Du Y, Deng Y, Fan H, Tao Y, Ma T (2021) Contribution of groundwater discharge and associated contaminants input to Dongting Lake, Central China, using multiple tracers (^{222}Rn, 18O, Cl^-). Environ Geochem Health 43:1239-1255

Swarzenski PW, Bratton JF, Crusius J (2004) Submarine ground-water discharge and its role in coastal processes and ecosystems. US Department of the Interior, US Geological Survey

Szymczycha B, Pempkowiak J (2015) The role of submarine groundwater discharge as material source to the Baltic Sea. Springer

Szymczycha B, Vogler S, Pempkowiak J (2012) Nutrient fluxes via submarine groundwater discharge to the Bay of Puck, southern Baltic Sea. Sci Total Environ 438:86–93

Szymczycha B, Borecka M, Białk-Bielińska A, Siedlewicz G, Pazdro K (2020) Submarine groundwater discharge as a source of pharmaceutical and caffeine residues in coastal ecosystem: Bay of Puck, southern Baltic Sea case study. Sci Total Environ 713:136522

Tamborski JJ, Cochran JK, Bokuniewicz HJ (2017) Submarine groundwater discharge driven nitrogen fluxes to Long Island Sound, NY: Terrestrial vs. marine sources. Geochim Cosmochim Acta 218:40–57

Taniguchi M, Dulai H, Burnett KM, Santos IR, Sugimoto R, Stieglitz T, Kim G, Moosdorf N, Burnett WC (2019) Submarine groundwater discharge: updates on its measurement techniques, geophysical drivers, magnitudes, and effects. Front Environ Sci 7:141

Taniguchi M (2002) Tidal effects on submarine groundwater discharge into the ocean. Geophys Res Lett 29(12):2–1

Thomas AT, von Harten J, Jusri T, Reiche S, Wellmann F (2022) An integrated modeling scheme for characterizing 3D hydrogeological heterogeneity of the New Jersey shelf. Mar Geophys Res 43(2):1–19

Urquidi-Gaume M, Santos IR, Lechuga-Deveze C (2016) Submarine groundwater discharge as a source of dissolved nutrients to an arid coastal embayment (La Paz, Mexico). Environ Earth Sci 75(2):1–13

Wang X, Li H, Yang J, Zheng C, Zhang Y, An A, Zhang M, Xiao K (2017) Nutrient inputs through submarine groundwater discharge in an embayment: a radon investigation in Daya Bay, China. J Hydrol 551:784–792

Wang Q, Li H, Zhang Y, Wang X, Zhang C, Xiao K, Qu W (2019a) Evaluations of submarine groundwater discharge and associated heavy metal fluxes in Bohai Bay, China. Sci Tot Environ 695:133873

Wang X, Li H, Zhang Y, Qu W, Schubert M (2019b) Submarine groundwater discharge revealed by ^{222}Rn: comparison of two continuous on-site ^{222}Rn-in-water measurement methods. Hydrogeol J 27(5):1879–1887

Wilson AM, Woodward GL, Savidge WB (2016) Using heat as a tracer to estimate the depth of rapid porewater advection below the sediment–water interface. J Hydrol 538:743–753

Wu Z, Zhou H, Zhang S, Liu Y (2013) Using ^{222}Rn to estimate submarine groundwater discharge (SGD) and the associated nutrient fluxes into Xiangshan Bay, East China Sea. Mar Pollut Bull 73(1):183–191

Xin P, Wang SS, Lu C, Robinson C, Li L (2015) Nonlinear interactions of waves and tides in a subterranean estuary. Geophys Res Lett 42(7):2277–2284

Young KS, Pradhanang SM (2021) Small unmanned aircraft (sUAS)-deployed thermal infrared (TIR) imaging for environmental surveys with implications in submarine groundwater discharge (SGD): methods, challenges, and novel opportunities. Remote Sens 13(7):1331

Yu S, Wang C, Li H, Zhang X, Wang X, Qu W (2022) Field and numerical investigations of wave effects on groundwater flow and salt transport in a sandy beach. Water Resour Res e2022WR032077

Zhang B, Zhang J, Yoshida T (2017) Temporal variations of groundwater tables and implications for submarine groundwater discharge: a 3-decade case study in central Japan. Hydrol Earth Syst Sci 21(7):3417–3425

Zhang Y, Wang X, Li H, Song D (2020) Grandes entradas de águas subterrâneas e fluxos associados de alcalinidade e nutrientes na Baía de Jiaozhou, China. Hydrogeol J 28:1721–1734

Zhang Y, Wang X, Xue Y, Zou C, Luo M, Li G, Li L, Cui L, Li H (2022) Advances in the study of submarine groundwater discharge (SGD) in China. Sci China Earth Sci 1–13

Chapter 7
Radon and Human Health

7.1 Introduction

This chapter's primary objective is to provide public health officials, medical professionals, toxicologists, and stakeholders a comprehensive understanding of the toxicology of radon. It includes insights on the applicability of toxicity and toxico-kinetic information to public health, wherever appropriate, and includes summaries and reviews of toxicological research and epidemiological investigations. This chapter is a prelude to the next chapter (Chap. 8) which addresses the mitigation measures to reduce health impacts of radon.

Soon after the proposal of air bubbling radium solutions to obtain radon gas by chemist Frederick Soddy in 1905, several physicians started recommending it as a panacea (Lavine 2013). Radon gained attention in the medical field for curing many diseases including tuberculosis, arthritis, asthma, sinusitis, etc. (Ranjit 2022). This paved the way for a new medicinal practice at spas, i.e., inhalation of radon collected in a specially designed enclosed room known as "inhalatorium" or "emanatorium", which exists even nowadays (Vogiannis and Nikolopoulos 2015).

Being an inert noble gas, radon does not interact chemically with other elements (Grandinetti 2018). Radioactive elements undergo spontaneous transformation/decay in which energy is emitted either in the form of particles (e.g., alpha and beta particles) or photons (e.g., gamma, X-rays). Progenies (daughter elements) are formed as a result of this decay process (Ntarisa et al. 2022). If these progenies are radioactive, they may undergo decay until reaching a stable state. The most commonly addressed isotopes of radon (^{219}Rn, ^{220}Rn, and ^{222}Rn) in nature are decay products of uranium (^{235}U, ^{238}U) and thorium (^{232}Th), and they end up as stable lead (Pb) towards the end of this decay chain (Lawal et al. 2021). ^{219}Rn is not considered in health impact studies, as their occurrence in environment is sparse. Also, the half-life of this isotope is very short (t1/2 = 4 s).

^{220}Rn (t1/2 = 56 s) is less preferred for toxicological studies, as the fraction of ^{220}Rn entering the environment is much less compared to ^{222}Rn despite its higher production rate. This is because of the slow diffusion of this isotope from ground

S. S. and S. Joseph, *Environmental Radon*, Environmental Science and Engineering,
https://doi.org/10.1007/978-981-99-2672-5_7

allows it to decay rapidly to form polonium which bonds with soil and a large proportion fails to reach atmosphere. Hence, ^{222}Rn with a considerable half-life (t1/2 = 3.8 days) is considered for radiological studies (Takahashi et al. 2022). From this point, the radon mentioned in all discussions of the text refers to ^{222}Rn unless otherwise specified.

7.2 Health Impacts of Radon

The health risks associated with radon are described in the following sub-sections.

(i) **Non-carcinogenic effects**

Besides its cancerous effects, a small proportion of studies claim that radon could pose non-carcinogenic effects, viz., chronic obstructive pulmonary disease (COPD), congenital malformations including the risk of cleft lip and palate, etc. (Turner et al. 2011; Langlois et al. 2016). Despite these findings, Turner (2012) later found that there is no positive association between radon and non-respiratory death cases. However, most of the investigations concluded that there is no correlation between radon activity and COPD (Kreuzer et al. 2013; Ruano-Ravina et al. 2021; Pando-Sandoval et al. 2022). In short, it is evident from the literature that the influence of radon on COPD incidence is still vivid (Barbosa-Lorenzo et al. 2017; Conde-Sampayo et al. 2020).

There is toxicological evidence showing premature development of atherosclerosis and increased risk of circulatory system diseases associated with external ionizing radiation exposure (Little et al. 2010; Kreuzer et al. 2015a, b). Based on this fact, Nusinovici et al. (2010) investigated the association of cumulative radon exposure with cerebrovascular diseases and coronary heart diseases in a cohort of uranium miners. The results suggested an association of radon with the former, but possible interrelation with cardiovascular risk factors was not possible to assess due to lack of data. Till date, knowledge on the interrelationship between radon exposure and its non-carcinogenic effects remains fragile.

(ii) **Carcinogenic effects**

Based on experimental evidence from studies on mutagenesis in lab animals and cell culture (Hussain et al. 1997; Weaver et al. 1997; Collier et al. 2005), as well as epidemiological cohort studies on uranium miners and case control studies on the general public (NRC 1999), radon is a known human lung carcinogen. The incidence of lung cancer is rising, making it the seventh most common cause of cancer-associated death worldwide among non-smokers (Kang et al. 2019, 2021). This is primarily caused by the frequent inhalation of radioactive ^{222}Rn gas and its decay products, such as polonium-218 (^{218}Po) and polonium-214 (^{214}Po), all of which are potent alpha particle emitters that bombard lung cells with alpha particle ionizing radiation (Gaskin et al. 2018; Sukanya et al. 2021). The International Agency for Research on Cancer has classified the alpha particle ionizing radiation from radon as a category-1 carcinogen because it damages lung cell DNA and causes genetic alterations that induce cancer (WHO 2009; Goodarzi et al. 2016; Samet 2020).

Besides its carcinogenic effects in never-smokers, recent findings suggest that ^{222}Rn is more potent in increasing lung cancer mortality among smokers (Lorenzo-González et al. 2019; Corrales et al. 2020). Because of its synergistic effects, radon-related lung cancer proneness in smokers is 25 times higher compared to non-smokers (Mc Laughlin et al. 2022). This health concern stimulated researchers to develop calculations of the radiation dose due to exposure in order to define the degree of risk and factors associated (Vogeltanz-Holm and Schwartz 2018; Fuente et al. 2019; Khan et al. 2019).

7.3 Measurement Units of Radioactivity

The rate of radioactivity or decay is generally expressed in curies (Ci) for radioactive elements. The activity refers to the number of radioactive transformations/disintegrations of a radionuclide over a given unit time. The curie is the fraction of radioactive materials in which 37 billion (3.7×10^{10}) disintegrations, i.e., decay events per each second.

For expressing radon, a much smaller unit, the picocurie (pCi) is used (pCi $= 1 \times 10^{-12}$ Ci). Becquerel (Bq) is an internationally accepted S.I. unit for expressing activity, i.e., the fraction of radioactive material in which one atom decays per second (1 Bq = ~27 pCi). Therefore, the activity concentration of any radionuclide like radon in air or water is expressed in pCi/L or Bq/m^3 of air. The product of its activity concentration and exposure time is termed as the exposure rate. Several models have been developed to estimate the radiation dose to tissue from this exposure (McColl et al. 2015; Zarnke et al. 2019). As the isotopes decay over a considerable time, some fraction of excretion occurs, the term dose is defined as the specific amount of radiant energy absorbed per unit mass in a specific tissue or organ which is expressed as rad (or gray).

7.4 Radon in Atmosphere

This section provides an extension to Sect. 1.3.2 of Chap. 1, which focused on the source, occurrence, and migration of radon in the atmosphere. The present section focuses specifically on the radiological pathways of radon in the atmosphere, offering a deeper understanding of their intricate properties and their implications for the environment and human health. By exploring these radiological pathways of radon in depth, this section provides readers with a comprehensive understanding of the behavior of radon and its progenies in the atmosphere.

Naturally occurring radioactive materials (NORMs) are ubiquitous in environment. The recent decades witnessed a lot of researches focusing on radon (^{222}Rn), the immediate progeny of ^{226}Ra, both are being the members of uranium (4n + 2) decay series (Boyle 2013; Mahmoudzadeh and Crawford 2023). Compared to its other isotopes, ^{222}Rn gained much prominence due to its isotopic abundance (99.27% wt.) in the total mixture of radon isotopes (Chao et al. 2020). ^{222}Rn is released from uranium-rich soil and a fraction (~10%) of it is evaded into atmosphere. ^{218}Po, ^{214}Pb, and ^{214}Bi are the most significant daughter products of ^{222}Rn from a radiological aspect (Rabi et al. 2017; Mostafa et al. 2020). Radon and these associated progenies interact indoors with aerosol particles as a result of complex physical interactions.

This interaction results in the generation of an inhalable radioactive mixture indoors. Two forms of radon progeny exist: aerosol-attached and unattached forms (Mostafa et al. 2020; Abdo et al. 2021). In most cases, radon progeny activity is associated with smaller particles with diameter spanning around 0.006–0.2 mm (mean = 0.025 mm). A small fraction of radon progeny (<0.1%) remains unattached and in dynamic equilibrium with attached fraction (Vogiannis and Nikolopoulos 2015). Atmospheric dust is associated with smaller unattached fraction and higher activities of radon due to additional radiation input from dust particles. Attached fraction settles down to earth by means of gravity and other processes. In soil, activity concentrations of ^{222}Rn are highly variable due to ^{226}Ra and ^{232}Th variations (Lara et al. 2015; Bangotra et al. 2018). In other words, atmospheric concentrations are greater over regions where ^{226}Ra and ^{232}Th in soil are abundant.

7.4.1 Factors Affecting ^{222}Rn in Air

Several factors influence the activity concentrations of ^{222}Rn in air: (i) ground variables (e.g., vegetation, building, paving, etc.), (ii) altitude above ground, (iii) geological variables (e.g., soil porosity and grain size of parent bed rock), (iv) atmospheric variables (e.g., barometric pressure, temperature, wind direction), (v) hydrological variables (precipitation, snow cover, soil moisture), and (vi) seasonal variables which depend on the geographical location (Mihailovic et al. 2014; Vogiannis and Nikolopoulos 2015; Cinelli et al. 2019; Dieu Souffit et al. 2022).

Generally, ground-level ^{222}Rn activity concentrations are maximum during autumn and early winter with a declining trend towards spring (Vogiannis and Nikolopoulous 2015; Xie et al. 2015). Later, Tchorz-Trzeciakiewicz et al. (2017) investigated the influence of terrain, geology, and weather conditions on atmospheric radon activity concentration above 900 m a.s.l in Kowary (SW Poland), a region characterized by temperate climate. They noticed that terrain and associated parameters, viz., inversion and weather stability conditions had significant impact on atmospheric ^{222}Rn activity concentration. On contrary to the general trend, ^{222}Rn activities in the region were found to be highest during summer and spring seasons, whereas lowest during autumn and winter due to inversion.

Reports on diurnal variation of ^{222}Rn show that its activities are highest during early morning due to temperature inversion. This trend gets reversed afternoon and again ^{222}Rn activities increase during sunset due to the cooling of earth and increased atmospheric stability (Taşköprü et al. 2023).

Correlations between ^{222}Rn activity and geophysical parameters have been observed in various studies. For instance, Ambrosino et al. (2020) found strong correlations between ^{222}Rn anomalies and fault displacements, fumarolic tremors, and earthquakes in Mladeč Caves, Czech Republic, and in soils of Phlegraean Fields caldera, Italy. The study investigated possible correlations between ^{222}Rn anomalies and fumarolic tremors in the Italian site, faults micro-displacements in the Czech site, as well as earthquakes that have an influence on the two studied areas. The results showed that the used method was highly effective, considering the calculated statistical uncertainties.

The outer temperature was identified as the main influencing radon driving force, and the extracted radon anomalies due to endogenous phenomena were well correlated with fault displacements, fumarolic tremors, and earthquakes under a characteristic delay time for each area. Additionally, significant correlations were found among earthquake magnitude and depth with fault displacement and fumarolic tremor.

7.5 Radon in Drinking Water

Radon in drinking water has been found to pose a carcinogenic risk, with 89% of the risk related to inhalation and 11% to ingestion of water containing the isotope (EPA 1991). A report from the National Academy of Science reveals that approximately 168 cancer deaths occur every year due to radon in drinking water (NRC 1999). Regulatory bodies such as the World Health Organization (WHO) and Environmental Protection Agency (EPA) have set maximum allowable levels of 100 and 11 Bq/L, respectively, for radon in drinking water (USEPA 1999; WHO 2004, 2022). Additionally, the Safe Drinking Water Act mandates the EPA to provide a higher alternative maximum contaminant level (AMCL) of 148 Bq/L with a multimedia mitigation program to tackle indoor air radon risks (USEPA 1999).

Meanwhile, Council Directive 2013/51/Euratom (2013) of the EU prescribes a range of parametric values between 100 and 1000 Bq/L for radon in drinking water across its member states (EURATOM 2013). The Directive mandates the specification of a parametric value within this range, with remedial action deemed necessary in all countries of the EU when the value exceeds 1000 Bq/L. Moreover, the World Health Organization (WHO) has set a reference dose level of 0.1 mSv per year for radon in drinking water, which is 10% of the annual dose limit of 1 mSv for the public (WHO 2004, 2011a).

It is imperative to adhere to these guidelines to ensure the safety of the general public from potential health hazards associated with ^{222}Rn in drinking water. Water sources should be treated to reduce radon levels well below the recommended threshold values. Before clearing new drinking water supplies for general consumption, radon testing should be conducted, and periodic testing may be necessary. The pathways through which ^{222}Rn enters buildings will be discussed in subsequent sections.

7.6 Radon Entry into Dwellings

Generally, ^{222}Rn in outdoor is not taken into account for health risk studies. This is because of the fact that outdoor ^{222}Rn is easily subjected to dilution. But once the ^{222}Rn diffuses into dwellings, the gas accumulates and becomes a health concern. Studies have demonstrated that the annual mean outdoor ^{222}Rn activity is about 10 Bq/m^3, while the annual mean indoor ^{222}Rn activity is around 40 Bq/m^3 (Čeliković et al. 2022). Moreover, the annual dose due to exposure to indoor ^{222}Rn is 80% higher (7000 h/yr occupancy time) than that of outdoor ^{222}Rn (Kuzmanović et al. 2020; Čeliković et al. 2022).

Understanding the pathways of ^{222}Rn into buildings is important for developing mitigation strategies. Radon emanates partially after its generation in soil-rock grains. This emanated radon migrates to short or long distances through existing pore spaces. Subsequently, it may enter into building structures via house pipes (e.g., for water and natural gas), or from building materials. The relative importance of these pathways depends on the local conditions (Collignan et al. 2012; Hess and Haskell 2020).

The possible entry points of radon into dwellings are depicted in Fig. 7.1. Specific consideration of the contributing sources is vital for establishing a radon mass balance for a building. Underlying geology (host rock, soil) and domestic water drawn from underground sources (groundwater) are more significant than building materials (Nero 2018).

A survey conducted on single-family houses in the United States has revealed alarming results regarding the emanation rates of concretes. The observed geometric mean emanation rate from these materials was found to be 20 Bq m^{-3} h^{-1}, which is significantly higher than the anticipated emission rates that were estimated to be in the range of 2–3 Bq m^{-3} h^{-1} (Ingersoll in 1983).

In another study, radon emanation rate from unattenuated soil (median = 25 Bq m^{-3} h^{-1}) demonstrated a close correspondence with indoor observations (Wilkening et al. 1972). Diffusion is the main mechanism favoring radon entry into the building from the ground.

7.7 Routes of Exposure

Radon and its progeny aggregately emit alpha, beta particles as well as gamma and X-rays during radioactive decay process. A major fraction of high-energy alpha particles emitted by radon progeny gets adhered to the trachea-bronchial tree and other parts of lungs through inhalation (Lawal et al. 2021; Ntarisa et al. 2022). The radioactive disintegration of ^{222}Rn to ^{218}Po is accompanied by emission of high-energy (5.5 meV) alpha particles, ^{218}Po to ^{214}Pb (6 meV) alpha particles and decay of ^{214}Bi and ^{214}Po releases long-range–low-energy beta and gamma radiation. Subsequently, ^{214}Po decays releasing high-energy (7.69 meV) alpha particles so rapidly (t1/2 = 1.6×10^{-4} s). This is followed by the disintegration of ^{214}Bi (t1/2

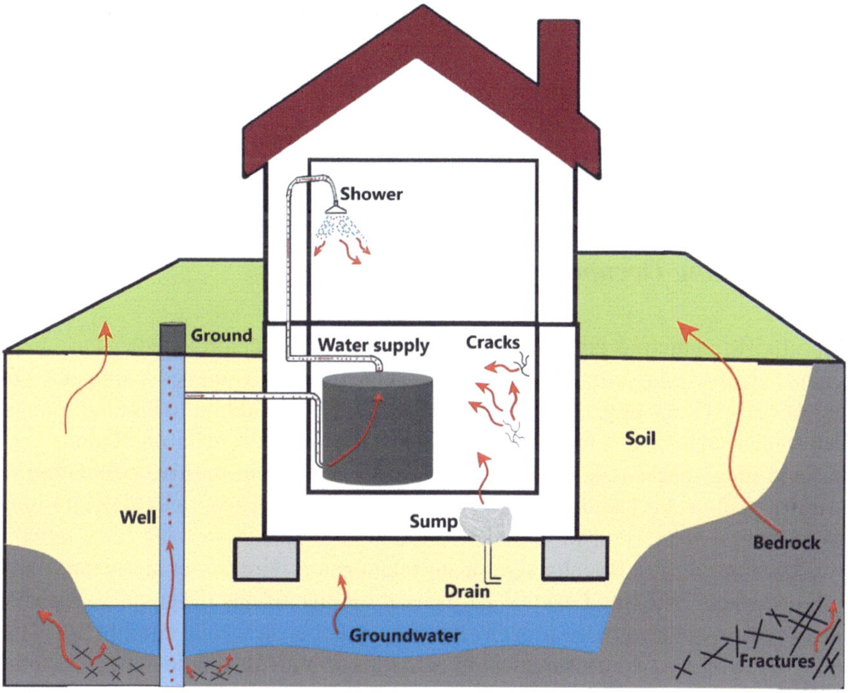

Fig. 7.1 Major entry points of radon into dwellings

= 19.7 min) and ^{214}Po involves release of beta and gamma radiations, which are of sufficiently long range–low energy to be considered of little relative toxicity concern to nearby cells (Madas et al. 2022).

Alpha emissions considered in radiation risk modeling are generally implied to ^{214}Bi decay (i.e., the decay rate of ^{214}Bi is essentially equal to the rate of ^{210}Pb produced due to the instantaneous decay of ^{214}Po from ^{214}Bi). Other radon progenies, viz., ^{210}Pb, ^{210}Bi, and ^{210}Po are not included in models as their contribution to radiation risk is negligible. For example, the half-life of ^{210}Pb is long enough (t1/2 = 22.2 years) and the natural biological clearance mechanisms limit the radiation dose associated with this isotope and other progenies. Henceforth, ^{218}Po and ^{214}Po are the radon progenies holding primary health concern especially in attached fraction.

When a radium atom disintegrates to form radon, the alpha particle is emitted. Radon and its progenies get attached to aerosol particles. These daughter isotopes readily aggregate, forming clusters and attach to the airborne dust particles as "attached fraction". This attached fraction when inhaled gets lodged in the tracheabronchial tree and starts bombarding adjacent cells repeatedly with alpha particles through the decay chain. This results in a large localized radiation dose inside the cells. The attached fraction is much higher in dwellings of smokers compared to homes of non-smokers and well-ventilated dusty mines (Grzywa-Celińska et al. 2020).

7.8 Radon in Working Environments

Most of the occupational radon exposure studies are confined to mining environments. In this section, reviews on radon exposure risks related to both mining and non-mining workplaces are presented.

7.8.1 Mining Occupational Exposure

Since 1950s, numerous epidemiological cohort and case control studies have been conducted among uranium miners to evaluate health risks from occupational exposure of radon (Rage et al. 2020; Madas et al. 2022). In addition to uranium mines, such studies were conducted in mining areas of fluorspar, tin, and iron. Radon exposure and lung cancer risk showed a statistically significant positive correlation in majority of the cohort studies. To be precise, the excess relative risk (ERR) for lung cancer mortality per unit of cumulative radon exposure in working level months (WLM) showed a statistically significant relationship (Kreuzer et al. 2015a, b). A term known as "Working Level (WL)" is used for quantifying the cumulative radiation dose from inhaled radon progeny which may not be in secular equilibrium. This unit was introduced as measurement of activities of individual radon progeny in the environment is not feasible during decay inside the body.

A recent study investigated the lung cancer mortality risk at low radon exposure rates in a sub-cohort ($n = 26{,}766$ miners) at a German uranium mining site (Kreuzer et al. 2015a, b). This study also provided strong evidence for the positive association of lung cancer mortality ($n = 334$ deaths) and cumulative exposure to radon in WLM, with an excess relative risk per WLM of 0.013 (95% confidence intervals).

The term Working Level (WL) refers to the amount of short-lived radon progeny concentration in one liter of air that emits 1.3×105 million electron volts (MeV) of alpha energy upon decay. The potential alpha energy of 2.08×10^{-5} J in one cubic meter of air (J/m^3) is equivalent to one WL. To convert the units of ^{222}Rn concentration from pCi/m^3 or Bq/m^3 to the potential alpha energy that its progeny can release upon complete decay (WL or J/m^3), the equilibrium between the radon gas and its progeny during exposure must be either known or assumed. When radon and its progeny reach a state of secular equilibrium, it is estimated that each pCi of radon present in the air will produce approximately 0.01 Working Level (WL), according to the EPA (2003) guidelines. In other words, 1 WL is equivalent to 100 pCi of radon. However, in indoor environments where removal processes such as air filtration or ventilation are employed, the activity concentration of short-lived progeny of radon may be lower than equilibrium levels.

In such cases, an equilibrium factor (F) is applied to account for this imbalance. For instance, the National Research Council Committee on Health Risks of Exposure to Radon (BEIR VI) assumes an equilibrium of 40% (i.e., $F = 0.4$) between radon and its progeny in residential settings (James et al. 2004). Under these guidelines, it is estimated that 1 pCi/L or 37 Bq/m^3 of ^{222}Rn in the air is equivalent to approximately 0.004 WL.

The Working Level Month (WLM) is the unit used to measure cumulative human exposure to radon progeny for mine workers. The WLM is calculated by multiplying the average concentration in Working Level (WL) by the duration of exposure in months. A working month is typically considered to be 170 h. The SI unit used for expressing WLM is J-hour/m^3, with 1 WLM equal to 3.6×10^{-3} J-hours/m^3 (Kreuzer et al. 2014; Kelly-Reif et al. 2020).

To convert from residential exposures expressed in pCi/L, it is assumed that on average, 70% of a person's time is spent indoors, and that 1 pCi/L of radon in indoor air is equivalent to 0.004 WL of radon progeny (Tomašek 2013; Nunes et al. 2022). This leads to the relationship: 1 pCi/L $\times$ 0.004 WL/pCi/L $\times$ 0.7 $\times$ (8,760 h/WL-year $\div$ 170 h/WL-M) = 0.144 WLM/year. Since 1 pCi/L is equivalent to 37 Bq/m^3, a residential exposure scenario using equivalent assumptions leads to the same cumulative exposure to radon progeny, i.e., 0.144 WLM/year (Chahine et al. 2011).

7.8.2 Non-mining Occupational Exposure

Research on radon-induced cancer risks among people dealing with non-mining work places, viz., schools, spas, show caves, etc. is very less in comparison with the number of works conducted in mining areas and residential dwellings. In thermal spas located in uranium-rich granitic regions, specific mitigatory actions are recommended which includes ventilation, systematic surveillance, monitoring, etc. (Silva and Dinis 2018).

Radon or progeny concentration measurements are performed for various purposes, including identifying workplaces with high levels of radon, conducting measurements in conjunction with remedial actions, and assessing worker exposure to comply with regulatory requirements (Martín Sánchez and Nuevo 2017; Bochicchio et al. 2022; Müllerová et al. 2022; Taşköprü et al. 2023). To identify the distribution of radon levels in workplaces, a large number of detectors are distributed, and cost and ease of placement are important factors to consider. The investigator must decide whether short-term or long-term measurements are more appropriate, taking into account potential seasonal influence such as heating, ventilation, and air conditioning (Appleton 2013; Altendorf et al. 2022). In most cases, long-term measurements over a period of several months are preferred due to their ability to capture seasonal variations in radon concentration (Lecomte et al. 2014). However, repeated measurements over shorter periods may be sufficient if adequate accuracy can be ensured.

When performing measurements for regulatory compliance, it is important to consider the frequency and duration of measurement, as well as the choice of instrumentation (see Chap. 2). This may involve monitoring the work area or using personal monitors worn by workers. The data collected will be used to assess individual doses and demonstrate compliance with regulatory requirements (Lecomte et al. 2014;

Gruber and Ringer 2017). If radon levels remain above the action level after remediation, a long-term measurement is necessary to confirm that the average radon level is below the action level. The decision on measurement frequency, duration, and instrumentation should be acceptable to the regulatory authority.

For estimating the annual effective dose due to indoor radon in air, the following equation (Ali et al. 2010) is used:

$$\mathrm{H}\left(\mathrm{mSv\,y^{-1}}\right) = \mathrm{CRn.\,F.\,O.\,(DCF)} \tag{7.1}$$

where C_{Rn} denotes the indoor radon activity concentration (Bq/m^3), F is the equilibrium factor between radon and its decay products (0.4), O is the average indoor occupancy time per person (in case of staff working at thermal spas, this occupancy time is approximately 2000 h per year based on the time spent at work), and DCF is the dose conversion factor for radon exposure, i.e., 7.9×10^{-6} mSv h^{-1} (Bq m^{-3})$^{-1}$. For school-based radon risk studies, an average indoor occupancy time per student or teacher can be much more lower (1620 h per year). A dose conversion factor of 9×10^{-6} mSv h^{-1} (Bq m^{-3})$^{-1}$ is usually considered (Lecomte et al. 2014).

Based on this equation, the annual effective doses due to radon exposure for staff working around the thermal pools in Niška Banja Spa (Serbia) were found to be very high (18 ± 1 mSv y^{-1}) than the ICRP65 recommendation level for workplaces, i.e., 2–10 mSv y^{-1} (Nikolov et al. 2012). A radon survey was conducted in 18 high schools in the radon-prone sites of Tenerife Island using 19 CR-39 passive detectors around a period of 90 days at a height of about 1–1.5 m above the floor representing the breathing height (López-Pérez et al. 2022). Continuous indoor radon measurements were carried out to investigate short-time (intra-day) fluctuations under different ventilation conditions. The annual dose contribution due to indoor radon exposure was found to be within ICRP recommendation level ranging from 0.07 to 1.18 mSv/year.

In a follow-up study on indoor radon levels in Portuguese thermal spas, Silva and Dinis (2022) conducted long-term measurements to estimate the year-round average radon exposure. The study involved 3 of the 16 previously studied thermal spas, where continuous measurements were taken for 1 year for indoor radon levels. The results showed that approximately 60% of the obtained values for radon concentration in the indoor air exceed the reference level of 300 Bq/m^3. The study included a time-series analysis of gamma dose rates that resulted in models capturing the dynamics of the data and providing short-term forecasts, offering valuable insights for developing sound planning and decision-making strategies in occupational exposure despite their inability to predict exact gamma dose rates.

7.9 Radon in Residential Dwellings

Radon is an ubiquitous air pollutant with low-concentration outdoors, but with potential for indoor buildup in poorly ventilated buildings. Human exposure to accumulated indoor radon levels is a significant contributor to lung cancer risk, with epidemiological evidence supporting the association between indoor radon exposure and increased risk of lung cancer (Darby et al. 2006; Zhang et al. 2012). The link between radon exposure and lung cancer risk has primarily been established through studies of miners exposed to high levels of radon gas in occupational settings, as compared to residential indoor radon exposure experienced by the general public.

7.9.1 Measurement Duration

Selecting the appropriate measurement duration based on the purpose and objective of a study is important while assessing the radiation dose due to radon. There are broadly two measurement types based on duration.

(i) **Long-Term Measurements**

Radon levels in dwellings are usually dynamic and significant temporal variation has been reported in numerous studies. On a diurnal scale, these variations occur by a factor of 2–3 within 24 h period. On a seasonal scale, even larger range of variations could be observed. Keeping this high variability on mind, long-term measurement is the most recommended one for obtaining more accurate and reliable annual average radon activity compared to short-term measurements.

Long-term measurements are typically carried out within a time frame of 3–12 months. During winter months, radon levels are usually high as the houses are sealed up. Once the measurement devices have been set at a particular location, no lifestyle change is required for the occupants during this type of measurement. However, it is recommended that the radon test should be performed in a home during heating season to understand the annual variation.

(ii) **Short-Term Measurements**

This type of measurement is not usually acceptable for assessing radon activity concentrations for planning remedial strategies. Due to the temporal variability of radon, it is strongly recommended that the results of any short-term measurement should be confirmed and validated using a follow-up long-term measurement, at the same location within the building for informing decisions on mitigation measures. However, short-term measurements serve the purpose when a rapid measurement of radon is required. For example, this method can be used for confirming and assessing the success rate of an already implemented mitigation strategy within the dwelling. In such situations, an initial short-term measurement (2–7 days) may be appropriate.

Assessing radon levels in homes can be conveniently done by measuring in a single location (Chung et al. 2020). For newly built homes, it is recommended to conduct radon measurements during the initial heating season after occupancy (Denman et al. 2013; Finne et al. 2019). Public Health England has recently advised that the radon activities in newly constructed homes within areas prone to high radon levels should be measured during the first year of occupancy (Denman et al. 2018). Homeowners must also contemplate re-testing whenever significant alterations or retrofitting are undertaken that could considerably alter ventilation or airflow in the home or change the usage of the lowest level inhabited rooms (Baeza et al. 2018; Reste et al. 2022). In such cases, a 3-month test should be performed during the initial heating season in newly refurbished dwellings to ensure safety (Chen and Moir 2012; Symonds et al. 2019; Gulan et al. 2022).

7.9.2 Measurement Location

All measurements should be taken in the typical occupancy area of the lowest occupied level of the dwelling in order to provide a reasonable assessment of radon exposure to the residents. The term "normal occupancy area" refers to any space used by a person for more than 4 hours each day. The following spaces could be used for measurements: living rooms, dens, playrooms, and bedrooms. While people spend the vast majority of their time in respective bedrooms rather than any other room in the house, low-level bedrooms (such as those on the main floor or basement) should be examined. Similar to this, facilities like basement playrooms should be examined if there are children present.

Despite its proneness to high radon activities, the basement is considered as a potential measurement site only if it is occupied for at least 4 h/day or if renovation plans are there. For example, adding a bedroom, playroom, or family room may result in an occupancy time more than 4 h/day. It is important to ensure that there will be no disturbance or interruption to the measurement device operation in the selected location throughout the measurement period.

The following criteria are followed during measurement:

- The most preferred location for placing the device is ideally near an interior wall at a height of 0.8–2 m from the floor, i.e., this height is the typical breathing zone for humans. There should be at least a 50 cm distance from ceiling and 20 cm from other objects to allow free movement of air around the device. This could be achieved by suspending the device, which again depends on the type of detector used. Considerable gaps should be maintained between the detector and walls (~40 cm for interior wall and ~50 cm in the case of exterior wall).
- Locations where the occupants do not spend considerable time (4 h/day) are not preferred, e.g., bathrooms.
- The measured radon levels should be a representative of the actual activity concentrations in the occupied area of the dwelling. Hence, locations near sump holes, closets, crawl spaces, etc. should be avoided for measurement.

- There should not be any presence of air currents generated by heating, ventilation, air conditioners, fans, windows, and doors.
- There is a possibility that the instrument could get affected if placed at locations exposed to heat. Hence, measurement devices should not be placed at locations exposed to direct sunlight, fireplaces, or radiators. This applies to the case of electrically powered equipment (computers, stereos, television, etc.) also.
- In dwellings lacking central air conditioning systems, radon measurements may be interrupted by open window conditions during warmer periods of the year. This uncertainty can be reduced significantly by practicing long-term radon measurement.

7.10 Radiation Dose Due to Radon in Water

Dissolved radon in water enters into dwellings through water distribution pipes. This dissolved radon gets released into indoor air when a faucet or valve is opened. This may occur, when the residents are taking a shower, cleaning the dishes or laundry. In most cases, a very small fraction of radon is added to indoor radon levels via outgassing. Radon levels are typically very low in municipally treated water systems. Depending on the source, radon levels in well water can be significant, although very high levels would be necessary to significantly lower indoor radon levels. It takes approximately 10,000 times the radon-in-water activity concentrations per m^3 of water to impact radon in air, i.e., indoor radon levels. Though such a level is a rare occurrence, it can happen occasionally in private or community wells.

When the indoor radon levels exceed the recommended 200 Bq/m^3 guideline, radon tests for groundwater (if the building depends on groundwater supply for water) should also be taken into consideration, especially if consistently high levels of groundwater radon are observed in the locality. Commercial test kits for radon in water are readily available for this purpose. In order to achieve an acceptable indoor radon level, it may be necessary to implement mitigation measures specifically designed to address radon originating from both soil and water sources.

In the section that follows, a case study from recent research illustrating the evaluation of radiation dosage owing to groundwater radon is presented.

7.10.1 Case Study—Karamana River Basin, India

A study was conducted by Sukanya et al. (2021) to investigate the distribution of radon (^{222}Rn) and associated radiation doses to the public in the Karamana River Basin (KRB), a small tropical basin located on the westward-facing slope of Southern Western Ghats of Kerala, India. The KRB is a vital water resource for Thiruvananthapuram City, and the study focused on evaluating the inhalation and ingestion doses of ^{222}Rn in groundwater from a health perspective. The study analyzed ^{222}Rn activities

Table. 7.1 ICRP age categories and corresponding water consumption rates by IAEA (1996)

Age category	Age in years	Water use (L/d)	Water use (L/y)
3 months	0–1	0.55	200
1 year	1–2	0.71	260
5 year	2–7	0.82	300
10 year	7–12	0.96	350
15 year	12–17	1.64	600
Adult	>17	2.00	730

in groundwater samples collected from 71 open wells located in highland, midland, and lowland regions of KRB.

The dose calculations were based on nuclides-specific (ICRP 1990) and age-specific (IAEA 1996) methodologies, and the annual effective dose (AED) and annual committed effective dose (ACED) were used to estimate the activity that enters the respiratory and gastrointestinal tracts from the environment. The ingested dose conversion factor of 5×10^{-9} Sv/Bq for ^{222}Rn nuclide (UNSCEAR 1993) and different age groups (IAEA 1996) were used for the calculations. The study evaluated the ACED received from ingestion of water containing radon for individual consumers under ICRP age groups using a specific equation (Table 7.1).

$$\text{Dose (Sv)} = {}^{222}\text{Rn activity concentration (Bq/L)} \times \text{annual water consumption (L)} \times \text{dose conversion factor (Sv/Bq)} \tag{7.2}$$

The annual mean effective dose resulting from both ingestion and inhalation was calculated based on the parameters introduced by UNSCEAR (UNSCEAR 2000), as outlined below:

$$\text{AED}_{\text{ing}}\left(\mu\text{Svy}^{-1}\right) = \text{Rn}^{222}\,\text{activity}\,\left(\text{BqL}^{-1}\right) \times \text{W}_{\text{in}} \times \text{DCF}_{\text{ing}} \tag{7.3}$$

$$\text{AED}_{\text{inh}}\left(\mu\text{Svy}^{-1}\right) = \text{Rn}^{222}\,\text{activity}\,\left(\text{BqL}^{-1}\right) \times \text{R}_{\text{aw}} \times \text{F} \times \text{O} \times \text{DCF}_{\text{inh}} \tag{7.4}$$

where AED_{ing} is the effective dose of ingestion (μSvy^{-1}), W_{in} is the ingestion intake rate (230, 330, and 730 Ly^{-1} for infants, children, and adults, respectively (WHO 2004, 2021), DCF_{ing} is the ingestion dose conversion factor (23, 5.9, and 3.5×10^{-9} Sv Bq^{-1} for infants, children, and adults, respectively (WHO 2004), AED_{inh} is the effective dose of inhalation (μSvy^{-1}), R_{aw} is the ratio of Rn in air to Rn in water (10^{-4}) (Gesell and Prichard 1980), F is the equilibrium factor between Rn and its decay products (0.4) (Chen and Harley 2018, 2020), O is the average indoor occupancy time per person (7000 hy^{-1}) (Martins et al. 2019; Sukanya et al. 2021; Ajiboye et al. 2022), and DCF_{inh} is the inhalation dose conversion factor (9nSv h^{-1} $(\text{Bq/m}^3)^{-1}$ (UNSCEAR 2000).

The statistical analysis revealed that the data had a highly variable, non-normal distribution, with a standard deviation greater than the median and mean values. The positive leptokurtosis (>3) and skewness of 3.6 indicated an asymmetric distribution of radon in the study area, which can be attributed to the variation in geological and structural formation. The radon activity varied significantly at different sampling sites, ranging from 0.17 to 68.3 BqL^{-1}, with a mean value of 7.1 BqL^{-1}.

All of the samples collected in the study were found to be within the acceptable parametric values for radon in drinking water set by the European Union (EU) in 2013 and the World Health Organization (WHO) in 2008, which is 100 BqL^{-1}. The transfer coefficient of radon from water to air is approximately 10^{-4}, meaning that for every 1 BqL^{-1} of radon in tap water, there is an increase of 0.1 Bqm^{-3} of radon in indoor air (WHO 2011b). Thus, tap water drawn from groundwater sources in the basin can release radon into indoor air in the range of 0.02–6.83 Bqm^{-3}. The study found that the mean annual dose due to inhalation of radon was about 12 times higher than the dose received from ingestion, indicating that respiratory epithelial cells are more susceptible to radon-induced risks than the gastric lining, which receives a negligible dose.

The estimated annual effective dose (AED) due to ingestion and inhalation of radon in groundwater showed a wide range from 0.5 to 208.4 μSvy-1 for infants, 0.4–172.2 μSvy^{-1} for children, and 0.5–189.7 μSvy^{-1} for adults, with mean values of 21.8, 19.4, and 19.8 μSvy^{-1}, respectively. These values are higher than the mean AED of 12.08 μSvy^{-1} reported in northern Rajasthan, India (Rani et al. 2013), but lower than the AED of 402 μSvy^{-1} found in the Sankey Tank Region of Karnataka, South India (Ravikumar and Somashekar 2014). Nonetheless, the mean AED values in the present study were well below the ICRP recommended action level of 3–10 $mSvy^{-1}$ for radon in dwellings (ICRP 1993).

The vulnerability analysis revealed that infants (<1 y) are the most sensitive group to radiation exposure due to radon, despite their lower water intake compared to children and adults. This is likely due to their low organ mass and rapidly dividing cells (Tubiana 2000; Stabin and Broga 2008; Alzen and Benz-Bohm 2011), which makes them more radiosensitive. The proliferative capacity of cells directly correlates with their susceptibility to DNA damage induced by radiation. Specifically, as the frequency of cell division increases, so does the likelihood of radiation-induced genotoxicity (Kesäniemi et al. 2019). In contrast, adults are the second most susceptible group to radon exposure due to their higher water intake, although the difference between adults and children was not significant. However, the estimated doses pose no considerable risk for any age group, according to the ICRP guidelines.

7.11 Conclusion and Future Research

For the past several years, scientists have been investigating the health impacts of indoor radon in dwellings. Radon is now a globally recognized potential lung carcinogen based on those epidemiological studies. Importance of this carcinogen is reflected in the initiatives, viz., global radon project by WHO with an objective to increase awareness, collect information, and motivate nations to take necessary actions in minimizing radon-related health hazards among public.

Although there are recommendation limits set by organizations like USEPA and WHO, there are no national regulations for ^{222}Rn in water and air in many countries including India. These nations still depend on an external benchmark values and use it as a reference. The radiation exposure to radon in dwellings depends on building structure, material, and location (local geological, meteorological conditions). Hence, there is a need to develop national/regional-level reference values for each country based on the local conditions.

More studies should concentrate on indoor radon distribution and its relationship with each unique construction style. Guidelines for standard limits should be downscaled from national to regional levels. National-level benchmark values could be used as a preliminary assessment tool, whereas regional-level values may be used for developing monitoring the performance of mitigation structures. Systematic national surveys should be carried out on a regular basis for mapping radon risk zones. Thus, early detection of this health concern could be attained and risk can be reduced.

References

Abdo MAS, Boukhair A, Fahad M, Ouakkas S, Arhouni FE, Hakkar M, Belahbib L, Al-Suhbani MN (2021) Estimation of unattached and aerosol-attached activities of airborne short-lived radon progeny in indoor environments. J Environ Radioact 237: 106665

Ajiboye Y, Isinkaye MO, Badmus GO, Faloye OT, Atoiki V (2022) Pilot groundwater radon mapping and the assessment of health risk from heavy metals in drinking water of southwest, Nigeria. Heliyon 8(2):e08840

Ali N, Khan EU, Akhter P, Khan F, Waheed A (2010) Estimation of mean annual effective dose through radon concentration in the water and indoor air of Islamabad and Murree. Radiat Prot Dosimetry 141(2):183–191

Altendorf D, Grünewald H, Liu TL, Dehnert J, Trabitzsch R, Weiß H (2022) Decentralised ventilation efficiency for indoor radon reduction considering different environmental parameters. Isot Environ Health Stud 58(2):195–213

Alzen G, Benz-Bohm G (2011) Radiation protection in pediatric radiology. Dtsch Arztebl Int 108(24):407

Ambrosino F, Thinová L, Briestenský M, Giudicepietro F, Roca V, Sabbarese C (2020) Analysis of geophysical and meteorological parameters influencing ^{222}Rn activity concentration in Mladeč caves (Czech Republic) and in soils of Phlegrean Fields caldera (Italy). Appl Radiat Isot 160:109140

Appleton JD (2013) Radon in air and water. In: essentials of medical geology. Springer, pp 239–277

Baeza A, García-Paniagua J, Guillén J, Montalbán B (2018) Influence of architectural style on indoor radon concentration in a radon prone area: a case study. Sci Total Environ 610:258–266

Bangotra P, Mehra R, Jakhu R, Kaur K, Pandit P, Kanse S (2018) Estimation of 222Rn exhalation rate and assessment of radiological risk from activity concentration of 226Ra, 232Th and 40K. J Geochem Explor 184:304–310

Barbosa-Lorenzo R, Ruano-Ravina A, Ramis R, Aragonés N, Kelsey KT, Carballeira-Roca C, Fernández-Villar A, López-Abente G, Barros-Dios JM (2017) Residential radon and COPD. An ecological study in Galicia, Spain. Int J Radiat Biol 93(2):222–230

Bochicchio F, Fenton D, Fonseca H, García-Talavera M, Jaunet P, Long S, Olsen B, Mrdakovic Popic J, Ringer W (2022) National radon action plans in Europe and need of effectiveness indicators: an overview of HERCA activities. Int J Environ Res Public Health 19(7):4114

Boyle RW (2013) Geochemical prospecting for thorium and uranium deposits. Elsevier

Čeliković I, Pantelić G, Vukanac I, Krneta Nikolić J, Živanović M, Cinelli G, Gruber V, Baumann S, Quindos Poncela LS, Rabago D (2022) Outdoor radon as a tool to estimate radon priority areas—a literature overview. Int J Environ Health Res 19(2):662

Chahine T, Schultz BD, Zartarian VG, Xue J, Subramanian SV, Levy JI (2011) Modeling joint exposures and health outcomes for cumulative risk assessment: the case of radon and smoking. Int J Environ Res Public Health 8(9):3688–3711

Chao JH, Ting CY, Huang FYJ, Tsai TL, Liu CC, Liu WC, Kang LC, Chin CY, Lin CC (2020) Background radiation in the production area of hokutolite in Taiwan. Radiat Phys Chem 172:108769

Chen J, Harley NH (2018) A review of indoor and outdoor radon equilibrium factors—Part I: ^{222}Rn. Health Phys 115(4):490–499

Chen J, Harley NH (2020) A review of radon equilibrium factors in underground mines, caves, and thermal spas. Health Phys 119(3):342–350

Chen J, Moir D (2012) The concept of equivalent radon concentration for practical consideration of indoor exposure to thoron. Int J Environ Res Public Health 9(1):286–293

Chung LK, Mata LA, Carmona MA, Shubayr NAM, Zhou Q, Ye Y, Kearfott KJ (2020) Radon kinetics in a natural indoor radon chamber. Sci Total Environ 734:139167

Cinelli G, De Cort M, Tollefsen T, Achatz M, Ajtić J, Ballabio C, Barnet I, Bochicchio F, Borelli P, Bossew P, Braga R (2019) Radon-Chapter 5. In: European atlas of natural radiation. Publication Office of the European Union

Collignan B, Lorkowski C, Améon R (2012) Development of a methodology to characterize radon entry in dwellings. Build Environ 57:176–183

Collier CG, Strong JC, Humphreys JA, Timpson N, Baker ST, Eldred T, Cobb L, Papworth D, Haylock R (2005) Carcinogenicity of radon/radon decay product inhalation in rats–effect of dose, dose rate and unattached fraction. Int J Radiat Biol 81(9):631–647

Conde-Sampayo A, Lorenzo-González M, Fernández-Villar A, Barros-Dios JM, Ruano-Ravina A (2020) Exposure to residential radon and COPD: a systematic review. Int J Chron Obstruct Pulmon Dis 15:939

Corrales L, Rosell R, Cardona A, Martín C, Zatarain-Barrón Z, Arrieta O (2020) Lung cancer in never smokers: the role of different risk factors other than tobacco smoking. Crit Rev Oncol Hematol 148:102895

Darby S, Hill D, Deo H, Auvinen A, Barros-Dios JM, Baysson H, Bochicchio F, Falk R, Farchi S, Figueiras A, Hakama M, Heid I, Hunter N, Kreienbrock L, Kreuzer M, Lagarde F, Mäkeläinen I, Muirhead C, Oberaigner W, Pershagen G, Ruosteenoja E, Rosario AS, Tirmarche M, Tomasek L, Whitley E, Wichmann HE, Doll R (2006) Residential radon and lung cancer—detailed results of a collaborative analysis of individual data on 7148 persons with lung cancer and 14208 persons without lung cancer from 13 epidemiologic studies in Europe. Scand J Work Environ Health 32(Suppl 1):1–83

Denman AR, Crockett RG, Groves-Kirkby CJ (2018) An assessment of the effectiveness of UK building regulations for new homes in Radon Affected Areas. J Environ Radioact 192:166–171

Denman AR, Sinclair J, Phillips PS, Crockett RG, Groves-Kirkby CJ (2013) The cost effectiveness of radon reduction programmes in domestic housing in England and Wales: the impact of improved radon mapping and housing trends. Environ Int 59:73–85

Dieu Souffit G, Jacob Valdes M, Bobbo Modibo O, Flore TSY, Ateba Jean Félix B, Tokonami S (2022) Radon risk assessment and correlation study of indoor radon, radium-226, and radon in soil at the cobalt–nickel bearing area of Lomié, Eastern Cameroon. Water Air Soil Pollut 233(6):196

EPA E (2003) EPA assessment of risks from radon in homes. Off Radiat Indoor Air, United States Environ Prot Agency, Washington, DC

EU (European Union) (2013) Laying down requirements for the protection of the health of the general public with regard to radioactive substances in water intended for human consumption. Council Directive 2013/51/Euratom of 22 October 2013

EURATOM (2013) Council directive on laying down requirements for the protection of health of general public with regard to radioactive substances in water intended for human consumption (2013/51/Euratom), Off J L296

Fuente M, Rábago D, Goggins J, Fuente I, Sainz C, Foley M (2019) Radon mitigation by soil depressurisation case study radon concentration and pressure field extension monitoring in a pilot house in Spain. Sci Total Environ 695:133746

Finne IE, Kolstad T, Larsson M, Olsen B, Prendergast J, Rudjord AL (2019) Significant reduction in indoor radon in newly built houses. J Environ Radioact 196:259–263

Gaskin J, Coyle D, Whyte J, Krewski D (2018) Utility gains from reductions in the modifiable burden of lung cancer attributable to residential radon in Canada. Can. J. Public Health 109:598–609

Gesell TF, Prichard HM (1980) The contribution of radon in tap water to indoor radon concentrations. In: Proceedings of natural radiation in the environment 2. NTIS, Houston, TX, pp 1347–1363

Grandinetti F (2018) Noble gas chemistry: structure, bonding, and gas-phase chemistry. John Wiley & Sons

Gruber V, Ringer W (2017) Radon exposure at specific workplaces in Austria—experiences and future challenges. Radiat Prot Dosimetry 177(1–2):7–11

Grzywa-Celińska A, Krusiński A, Mazur, J., Szewczyk K, Kozak K (2020) Radon—the element of risk. The impact of radon exposure on human health. Toxics 8(4):120

Goodarzi AA, Anikin A, Pearson DD (2016) Environmental sources of ionizing radiation and their health consequences. In Genome Stability. 569–581 Academic Press, United Kingdom

Gulan L, Stajic JM, Spasic D, Forkapic S (2022) Radon levels and indoor air quality after application of thermal retrofit measures—a case study. Air Qual Atmos Health 1–11

Hess CT, Haskell LE (2020) A comparison of two pathways to human for radon from drinking water: inhalation and ingestion. In: Water contamination and health. CRC Press, pp 167–198

Hussain SP, Kennedy CH, Amstad P, Lui H, Lechner JF, Harris CC (1997) Radon and lung carcinogenesis: mutability of p53 codons 249 and 250 to ^{238}Pu alpha-particles in human bronchial epithelial cells. Carcinogenesis 18(1):121–125

IAEA (1996) International basic safety standards for protection against ionizing radiation and for the safety of radiation Sources. IAEA Saf Ser 115. International Atomic Energy Agency, Vienna

ICRP (International Commission on Radiological Protection) (1990) Recommendations of the international commission on radiological protection. Annals of the ICRP 21(1–3). ICRP Publication; 60. Pergamon Press, Oxford

ICRP (International Commission on Radiological Protection) (1993) Protection against radon-^{222} at home and at work. Int Comm Radiol Protect Ann ICRP 23(2):65. ICRP Publication

Ingersoll JG (1983) A survey of radionuclide contents and radon emanation rates in building materials used in the US. Health Phys 45(2):363–368

James AC, Birchall A, Akabani G (2004) Comparative dosimetry of BEIR VI revisited. Radiat Prot Dosimetry 108(1):3–26

Kang JK, Seo S, Jin YW (2019) Health effects of radon exposure. Yonsei Med J 60(7):597–603

Kang J, Singh BK, Um W (2021) Efficient radon removal using fluorine-functionalized natural zeolite. J Environ Radioact 233:106607

Kelly-Reif K, Sandler DP, Shore D, Schubauer MK, Troester MA, Nylander-French L, Richardson DB (2020) Radon and cancer mortality among underground uranium miners in the Příbram region of the Czech Republic. Am J Ind Med 63(10):859–867

Kesäniemi J, Lavrinienko A, Tukalenko E, Boratyński Z, Kivisaari K, Mappes T, Milinevsky G, Møller AP, Mousseau TA, Watts PC (2019) Exposure to environmental radionuclides associates with tissue-specific impacts on telomerase expression and telomere length. Sci Rep 9(1):1–9

Khan SM, Krewski D, Gomes J, Deonandan R (2019) Radon, an invisible killer in Canadian homes: perceptions of Ottawa-Gatineau residents. Can J Public Health 110:139–148

Kreuzer M, Auvinen A, Cardis E, Hall J, Jourdain JR, Laurier D, Little MP, Peters A, Raj K, Russell NS, Tapio S (2015a) Low-dose ionising radiation and cardiovascular diseases—strategies for molecular epidemiological studies in Europe. Mutat Res/Rev Mutat Res 764:90–100
Kreuzer M, Fenske N, Schnelzer M, Walsh L (2015b) Lung cancer risk at low radon exposure rates in German uranium miners. Br J Cancer 113(9):1367–1369
Kreuzer M, Dufey F, Marsh JW, Nowak D, Schnelzer M, Walsh L (2014) Mortality from cancers of the extra-thoracic airways in relation to radon progeny in the Wismut cohort, 1946–2008. Int J Radiat Biol 90(11):1030–1035
Kreuzer M, Sogl M, Brüske I, Möhner M, Nowak D, Schnelzer M, Walsh L (2013) Silica dust, radon and death from non-malignant respiratory diseases in German uranium miners. Occup Environ Med 70(12):869–875
Kuzmanović P, Todorović N, Filipović Petrović L, Mrđa D, Forkapić S, Nikolov J, Knežević J (2020) Radioactivity of building materials in Serbia and assessment of radiological hazard of gamma radiation and radon exhalation. J Radioanal Nucl Chem 324:1077–1087
Langlois PH, Lee M, Lupo PJ, Rahbar MH, Cortez RK (2016) Residential radon and birth defects: A population-based assessment. Birth Defects Res A Clin Mol Teratol 106(1):5–15
Lara E, Rocha Z, Palmieri HEL, Santos TO, Rios FJ, Oliveira AH (2015) Radon concentration in soil gas and its correlations with pedologies, permeabilities and 226Ra content in the soil of the Metropolitan Region of Belo Horizonte–RMBH, Brazil. Radiat Phys Chem 116:317–320
Lavine M (2013) The first atomic age: scientists, radiations, and the American public, 1895–1945. Springer, New York
Lawal MK, Ayanlola PS, Oladapo OO, Oni MO, Aremu AA (2021) Assessment of radon exhalation rates in mineral rocks used in building decoration in Nigeria. Radiat Prot Environ 44(3):161
Lecomte JF, Solomon S, Takala J, Jung T, Strand P, Murith C, Kiselev S, Zhuo W, Shannoun F, Janssens A (2014) ICRP publication 126: radiological protection against radon exposure. Ann ICRP 43(3):5–73
Little MP, Tawn EJ, Tzoulaki I, Wakeford R, Hildebrandt G, Paris F, Tapio S, Elliott P (2010) Review and meta-analysis of epidemiological associations between low/moderate doses of ionizing radiation and circulatory disease risks, and their possible mechanisms. Radiat Environ Biophys 49(2):139–153
López-Pérez M, Hernández F, Díaz JP, Salazar-Carballo PA (2022) Determination of the indoor radon concentration in schools of Tenerife (Canary Islands): a comparative study. Air Qual Atmos Health 15(5):825–835
Lorenzo-González M, Torres-Durán M, Barbosa-Lorenzo R, Provencio-Pulla M, Barros-Dios JM, Ruano-Ravina A (2019) Radon exposure: a major cause of lung cancer. Expert Rev Respir Med 13:839–850
Madas BG, Boei J, Fenske N, Hofmann W, Mezquita L (2022) Effects of spatial variation in dose delivery: what can we learn from radon-related lung cancer studies? Radiat Environ Biophys 1–17
Mahmoudzadeh B, Crawford J (2023) Modelling simultaneous transport of natural and anthropogenic radionuclides in fractured media—diffusion into the heterogeneous layered rock matrix for an arbitrary length decay chain. J Contam Hydrol 104138
Martín Sánchez A, Nuevo MJ (2017) Actions for remediation in cases with large concentration of radon indoor. J Radioanal Nucl Chem 311:1219–1225
Martins L, Pereira A, Oliveira A, Sanches LF, Pacheco FA (2019) A new framework for the management and radiological protection of groundwater resources: the implementation of a Portuguese action plan for radon in drinking water and impacts on human health. Water 11(4):760
McColl N, Auvinen A, Kesminiene A, Espina C, Erdmann F, de Vries E, Greinert R, Harrison J, Schüz J (2015) European code against Cancer 4th edition: Ionising and non-ionising radiation and cancer. Cancer Epidemiol 39(Suppl 1):S93–S100
Mc Laughlin JP, Gutierrez-Villanueva JL, Perko T (2022) Suggestions for improvements in national radon control strategies of member states which were developed as a requirement of EU directive 2013/59 EURATOM. Int J Environ Res Public Health 19(7):3805

Mihailovic DT, Udovičić V, Krmar M, Arsenić I (2014) A complexity measure based method for studying the dependance of 222Rn concentration time series on indoor air temperature and humidity. Appl Radiat Isot 84:27–32

Mostafa MYA, Khalaf HNB, Zhukovsky M (2020) Radon decay products equilibrium at different aerosol concentrations. Appl Radiat Isot 156:108981

Müllerová M, Mrusková L, Holý K, Smetanová I, Brandýsová A (2022) Estimation of seasonal correction factor for indoor radon concentration in Slovakia: a preliminary survey. J Radioanal Nucl Chem 1–6

Nero, A.V., 2018. Indoor concentrations of radon-^{222} and its daughters: sources, range, and environmental influences. In: Indoor air and human health). CRC Press, pp 43–68

Nikolov J, Todorovic N, Petrovic Pantic T, Forkapic S, Mrdja D, Bikit I, Krmar M, Veskovic M (2012) Exposure to radon in the radon spa Niška Banja, Serbia. Radiat Meas 47:443–450

NRC (National Research Council) (1999) Risk assessment of radon in drinking water. Committee on Risk Assessment of Exposure to Radon in Drinking Water. ISBN:0-309-52474-1. National Academies Press, p 296

Ntarisa AV, Kim HJ, Aryal P, Quang ND, Saha S (2022) Novel technique for simultaneous detecting three naturally occurring radon isotopes (^{222}Rn, 220Rn and 219Rn). Radiat Phys Chem 110202

Nunes LJ, Curado A, Graça LCD, Soares S, Lopes SI (2022) Impacts of indoor radon on health: a comprehensive review on causes, assessment and remediation strategies. Int J Environ Res Public Health 19(7):3929

Nusinovici S, Vacquier B, Leuraud K, Metz-Flamant C, Caër-Lorho S, Acker A, Laurier D (2010) Mortality from circulatory system diseases and low-level radon exposure in the French cohort study of uranium miners, 1946–1999. Scand J Work, Environ Health 373–383

Pando-Sandoval A, Ruano-Ravina A, Torres-Durán M, Dacal-Quintas R, Valdés-Cuadrado L, Hernández-Hernández JR, Consuegra-Vanegas A, Candal-Pedreira C, Varela-Lema L, Fernández-Villar A, Pérez-Ríos M (2022) Residential radon and characteristics of chronic obstructive pulmonary disease. Sci Rep 12(1):1–7

Rabi R, Oufni L, Amrane M (2017) Modeling of indoor 222Rn distribution in ventilated room and resulting radiation doses measured in the respiratory tract. J Radiat Res Appl Sci 10(3):273–282

Rage E, Richardson DB, Demers PA, Do M, Fenske N, Kreuzer M, Samet J, Wiggins C, Schubauer-Berigan MK, Kelly-Reif K, Tomasek L (2020) PUMA–pooled uranium miners analysis: cohort profile. Occup Environ Med 77(3):194–200

Rani A, Mehra R, Duggal V (2013) Radon monitoring in groundwater samples from some areas of northern Rajasthan, India, using a RAD7 detector. Radiat Prot Dosim 153(4):496–501

Ranjit M (2022) Hot and mineral spring water for health benefits. In: Hot Springs in Nepal. Springer, Cham, pp 161–187

Ravikumar P, Somashekar RK (2014) Determination of the radiation dose due to radon ingestion and inhalation. Int J Environ Sci Technol 11(2):493–508

Reste J, Pavlovska I, Martinsone Z, Romans A, Martinsone I, Vanadzins I (2022) Indoor air radon concentration in premises of public companies and workplaces in Latvia. Int J Environ Res Public Health 19(4):1993

Ruano-Ravina A, Cameselle-Lago C, Torres-Durán M, Pando-Sandoval A, Dacal-Quintas R, Valdés-Cuadrado L, Hernández-Hernández J, Consuegra-Vanegas A, Tenes-Mayén JA, Varela-Lema L, Fernández-Villar A (2021) Indoor radon exposure and COPD, synergic association? a multicentric, hospital-based case–control study in a radon-prone area Archivos De Bronconeumología (english Edition) 57(10)630–636

Samet JM (2020) Carcinogenesis and lung cancer: 70 years of progress and more to come. Carcinogenesis 41(10):1309–1317

Silva AS, Dinis MDL (2022) Assessment of indoor radon concentration and time-series analysis of gamma dose rate in three thermal spas from Portugal. Environ Monit Assess 194(9):611

Silva AS, Dinis ML (2018) Main mitigation measures—occupational exposure to radon in thermal spas. In: Occupational safety and hygiene VI. CRC Press, pp 425–429

Stabin MG, Broga D (2008) Radiation protection and dosimetry: an introduction to health physics. Med Phys 35(10):4766

Sukanya S, Joseph S, Noble J (2021) Evaluation of radiation dose from radon ingestion and inhalation in groundwater of a small tropical river basin, Kerala, India. Isot Environ Health Stud 57(2):204–215

Symonds P, Rees D, Daraktchieva Z, McColl N, Bradley J, Hamilton I, Davies M (2019) Home energy efficiency and radon: An observational study. Indoor Air 29(5):854–864

Takahashi LC, de Oliveira Santos T, Pinheiro RMM, Passos RG, Rocha Z (2022) Radon dosimetry using Solid State Nuclear Track Detectors in different environments: a review. Appl Radiat Isot 186:110217

Taşköprü C, İçhedef M, Saç MM (2023) Diurnal, monthly, and seasonal variations of indoor radon concentrations concerning meteorological parameters. Environ Monit Assess 195(1):25

Tchorz-Trzeciakiewicz DE, Klos M (2017) Factors affecting atmospheric radon concentration, human health. Sci Total Environ 584:911–920

Tomašek L (2013) National Radiation Protection Institute. Fundam Assess Risks Environ Radiat 55:479

Tubiana M (2000) Radiation risks in perspective: radiation-induced cancer among cancer risks. Radiat Environ Biophys 39(1):3–16

Turner MC, Krewski D, Chen Y, Pope CA, Gapstur S, Thun MJ (2011) Radon and lung cancer in the American Cancer Society cohort. Cancer Epidemiol Biomark Prev 20(3):438–48

UNSCEAR (1993) Sources and effects of ionizing radiation. UNSCEAR 1993 report to general assembly, with scientific annexes, 18. United Nations Scientific Committee on the Effects of Radiation (UNSCEAR), New York

UNSCEAR (2000) Sources and effects of ionizing radiation. Report to the general assembly, with scientific annexes, vol 1: sources. United Nations Scientific Committee on the Effects of Atomic Radiation, New York (NY)

USEPA (US Environmental Protection Agency) (1999) National primary drinking water regulations; Radon-222: Proposed Rule, 59246–59378

USEPA (1991) National primary drinking water regulations; Radionuclides; Proposed Rule. 40 CFR. United States Environmental Protection Agency, Washington (DC)

Vogeltanz-Holm N, Schwartz GG (2018) Radon and lung cancer: what does the public really know? J Environ Radioact 192:26–23

Vogiannis EG, Nikolopoulos D (2015) Radon sources and associated risk in terms of exposure and dose. Front Public Health 2:1–10

Weaver DA, Hei TK, Hukku B, McRaven JA, Willey JC (1997) Cytogenetic and molecular genetic analysis of tumorigenic human bronchial epithelial cells induced by radon alpha particles. Carcinogenesis 18:1251–1257

WHO (2004) Guidelines for drinking-water quality, vol 1. World Health Organization, Geneva

WHO (World Health Organization) (2004a) Guidelines for drinking water quality. Health criteria and other supporting information, vol 1. World Health Organization, Geneva

WHO (World Health Organization) (2004b) Guidelines for drinking-water quality, vol 1, 3rd edn. World Health Organization

WHO (World Health Organization) (2008) Guidelines for drinking water quality. Incorporating first and second Addenda, 3rd edn. WHO Press, Geneva

WHO, World Health Organization (2009) Handbook on indoor radon: a public health perspective. WHO, Geneva

WHO (World Health Organization) (2011a) Guidelines for drinking-water quality. World Health Organization 216:303–304

WHO (World Health Organization) (2011b) Guidelines for drinking water quality. Chapter 9, Radiological aspects, 4th edn. WHO Press, Geneva

WHO (2021) A global overview of national regulations and standards for drinking-water quality. World Health Organization, Geneva

WHO (World Health Organization) (2022) Guidelines for drinking-water quality: incorporating the first and second addenda. World Health Organization, Geneva

Wilkening MH, Clements WE, Stanley D (1972) Radon ^{222} flux measurements in widely separated regions (No. CONF-720805--P2)

Xie D, Liao M, Kearfott KJ. Influence of environmental factors on indoor radon concentration levels in the basement and ground floor of a building–a case study. Radiat Meas. 2015;82:52–8

Zarnke AM, Tharmalingam S, Boreham DR, Brooks AL (2019) BEIR VI radon: the rest of the story. Chem Biol Interact 301:81–87

Zhang Z-L, Sun J, Dong J-Y, Tian H-L, Xue L, Qin L-Q, Tong J (2012) Residential radon and lung cancer risk: an updated meta-analysis of case-control studies. Asian Pac J Cancer Prev 13(6):2459–2465

Chapter 8
Radon—Mitigatory and Control Measures

8.1 Introduction

Radon has been recognized as a major threat to human health for over 50 years, with an estimated 20% of global lung cancer deaths attributed to radon exposure (Kim et al. 2016). Despite this knowledge, very few countries have taken statutory action to address this issue. As a result, there is a pressing need to implement effective mitigation and control measures to reduce the risk of indoor radon exposure (Vogeltanz-Holm and Schwartz 2018; Khan et al. 2019).

Even though the risk posed by radon to human health has been recognized for more than 50 years and has been linked to around 20% of lung cancer deaths worldwide (Kim et al. 2016), barely a few countries have lately passed statutory action to address the issue. Under such scenarios, there has been a growing need to implement mitigation measures to reduce the risk of indoor radon exposure (Vogeltanz-Holm and Schwartz 2018; Khan et al. 2019).

For several decades, recommendation and guidelines on controlling radon exposure have been provided by numerous international bodies viz., United States Environment Protection Agency (USEPA), International Commission on Radiological Protection (ICRP), International Atomic Energy Agency (IAEA), World Health Organization (WHO) and European Union (EU). An efficient radon control technology should be incorporated during every new construction, especially in places where radon levels are commonly reported to be high, according to the International Residential Code (George 2015). With an objective to inspect radon in workplace and protect individuals from radon risk across the European Union, the European Directive 2013/59/Euratom- Basic Safety Standards Directive was implemented in 2013 (Silva and Dinis 2021). Several states in the USA and most of Europe have deployed building codes that adhere to global standards (Stanley et al. 2019).

Canada is also one of the pioneering countries in implementing well established guidelines for radon in indoor environment. Canadian federal government has customized its own model national building code allocating a broad policy direction with no legal bindings (Gaskin et al. 2021). In 2017, most of the territories and

S. S. and S. Joseph, *Environmental Radon*, Environmental Science and Engineering,
https://doi.org/10.1007/978-981-99-2672-5_8

provinces under Canadian government amended the building codes in order to reduce radon exposure in newly constructed buildings (Khan et al. 2021).

8.2 Mitigation Strategies

The radon mitigation technologies developed so far are technically feasible for reducing indoor radon levels below reference levels. The regular drafting and modification of building codes increases the scope for controlling radon levels prior to the construction of new buildings and restructuring the existing buildings based on the radon activity measurements (Bastian et al. 2022; Mc Laughlin et al. 2022).

While a few national building codes incorporate radon safeguards as design and construction criteria for future dwellings, the majority of these codes do not include a requirement for a post-construction assessment to measure the radon level in the completed structure. Majority of the countries do not mandate homeowners of existing homes to conduct radon testing, mitigate (i.e., reduce radon levels using temporary alterable methods), and remediate (i.e., install more systematic sustainable techniques to reduce radon levels below the reference level) (Cori et al. 2022).

Despite the availability of updated building codes and accessible free handbooks and guidelines published online, there are no strict rules for checking the radon status while buying or selling a property through real estate dealings. This gap is reflected in dwellings of Alberta (Canada) where the mean radon levels were around 31.5% higher in new constructions than existing building structures before early 1990s (Stanley et al. 2017). This awakened the importance of radon mitigation systems among the population. Consequently, this lead to the implementation of building codes for different types of dwellings.

Radon control strategies with mitigation alone are incomplete when preventive approaches are lacking in the framework. The chief components of a successful national radon programme are as follows:

1. Designing radon control strategies should be compatible for both new and existing structures.
2. Strategies should be cost-effective and affordable for all economic classes.
3. Region specific building codes and mitigation measures should be developed, as radon sources are highly variable between regions.
4. Mechanisms involving pressure-driven soil gas infiltration, groundwater radon migration, emanation from building materials etc. should be taken into consideration.
5. Appropriate training and certification of building professionals should be implemented.

8.3 Design Criteria for Radon Control Systems

The following conditions should be taken into account while designing radon control (preventive and mitigation) systems:

- High efficiency to considerably reduce radon activity concentrations below the reference level;
- Highly durable and functional for a long term period;
- Easy to install any additional structures (e.g., inline fan) in future
- Should prevent back-drafting;
- Monitoring should be simple and easy;
- Cost effective, easy to operate and maintain.

8.4 Guidelines, Standards and Regulatory Bodies

Effective control of radon exposure requires the collaboration of experts from various fields, including radon mitigation contractors, building planners, and researchers. The development of appropriate guidelines and standards is crucial in ensuring the safety of people from the harmful effects of radon. Several countries, including Austria, Belgium, China, Czech Republic, Finland, France, Ireland, Latvia, Norway, Russia, Sweden, Switzerland, the United Kingdom, and the United States of America, have already implemented mitigation or prevention guidance documents and standards (WHO 2007).

Constant efforts have been made worldwide to prevent and reduce the degree of radon exposure in new buildings. Table 8.1 portrays some of the regulatory bodies in different geographical regions authorized for preventing ^{222}Rn entry in new buildings.

The Czech Republic has implemented regulatory measures that require radon testing at every new building site prior to construction. This testing establishes a

Table 8.1 Regulatory bodies/authorities for prevention of ^{222}Rn in new buildings

Region	Regulatory body/authority
Canada	Canadian Commission on Building and Fire Codes
Ireland	Department of Environment, NSAI/Irish Agreement Board
Czech Republic	State Office for Nuclear Safety; Czech Office for Standards
Germany	Ministry of Interior of Saxony; Federal Ministry of Environment; Nature Conservation and Nuclear Safety, Ministries of Environment of the German Federal Lands
Denmark	Ministry of Housing
Poland	Polish Atomic Energy Agency
Finland	Building Information Ltd. and Ministry of Environment
Sweden	National Board of Housing Building and Planning

radon index for the site, which is then used to determine the appropriate level of radon prevention required for construction (Gruber et al. 2013). In other parts of the world, such as the United States of America and several countries in Europe, cost-effective radon control systems are installed in all new buildings to mitigate radon exposure (NRPA 2013; Finne et al. 2019; Burghele et al. 2021; Mc Laughlin et al. 2022; Alabid et al. 2022).

In 2009, the government of Norway adopted a national strategy to minimize radon exposure (Norwegian Ministries 2010). This program included a two-pronged approach: (i) lowering radon levels in all buildings below standard limits, and (ii) minimizing radon exposure as low as reasonably achievable (ALARA). This program introduced mandatory preventive measures during the construction phase of new buildings and legally binding limits on radon levels. The aim was to ensure that radon levels in buildings are reduced to safe limits, thereby minimizing the risk of lung cancer due to radon exposure.

8.5 Radon Prevention Strategies for New Constructions

The significant radon transport mechanism is pressure-driven airflow from the soil to the dwelling, and diffusion is also one of the driving forces (Wang et al. 2022). Radon penetration mainly occurs due to air pressure differences between the soil and the occupied dwelling, and thus radon prevention measures usually focus on reversing this pressure differential (Fuente et al. 2019; Goggins et al. 2019). In new constructions, low indoor radon activity concentrations are challenging to achieve and sustain due to radon management measures. Therefore, radon testing during various occupancy stages is recommended for new constructions during different stages of occupancy:

- **Pre-occupancy phase**: Indoor radon activity concentrations in an unoccupied building may differ than in occupied dwelling due to the heating and ventilation differences (Akbari et al. 2013; Ivanova et al. 2017). However, radon testing prior to occupancy in a new building can be helpful in identifying problems. This way, it may be easier to rectify exposure related issues at this early stage than during occupancy.
- **Occupancy phase**: During the occupancy phase, radon measurements have to be carried out to confirm whether indoor radon activity concentrations are below the reference level (Gulan et al. 2022). Radon testing should be done in regular intervals to ensure the performance of radon control systems in the building.

Various techniques have been developed and implemented to minimize indoor radon levels in new buildings. Passive soil depressurization (PSD), active soil depressurization (ASD), surface sealing, membranes, ventilation of unoccupied spaces, and ventilation of occupied spaces are among the most commonly applied strategies (Jiránek and Kačmaříková 2014; Goggins et al. 2019; Ruvira et al. 2022). The majority of these prevention strategies rely on the principle of air pressure differences between the soil and indoor occupied space. In certain cases, a combination of approaches may be necessary, particularly in buildings with various types of foundations.

Several prevention strategies are summarized and listed in the following subsections:

a. Passive soil depressurization (PSD)

Passive soil depressurization (PSD) systems typically include a suction point, such as a sump or a vertical vent pipe, installed ideally in a uniform permeable aggregate layer beneath the building foundation. The pressure differences generated by wind blowing across the vent pipe exit point create a natural stack effect, which draws soil gases upward and out of the building (Fuente et al. 2019; Zhou et al. 2021). Unlike active soil depressurization systems that require fans to create suction, PSD systems do not require any additional components. The terminal of the vent pipe can be left open or connected to a cowl, such as a static or rotating or ridge vents. Among these, rotating cowls have demonstrated the best performance in reducing radon levels, with a reduction factor of up to 2.1 (Hodgson and Pudner 2019a, b). An example of a PSD system and its basic components can be seen in Fig. 8.1.

The temperature differences between indoor and outdoor environments in a building generate a pressure gradient due to the stack effect, resulting in an increase of soil gas flow from the ground into the basement through the existing entry points (Frutos et al. 2021). Pressure gradient due to stack effect can be calculated using the

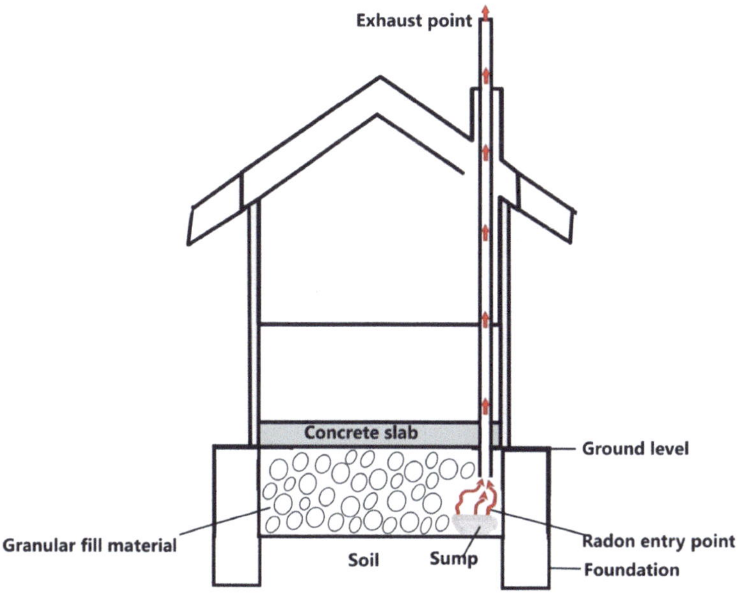

Fig. 8.1 Passive soil depressurization for radon control in new constructions

following equation (Fuente et al. 2019):

$$\Delta P = cP_{atm}h\left(\frac{1}{T_{out}} - \frac{1}{T_{in}}\right) \tag{8.1}$$

where ΔP is the indoor-outdoor pressure difference, $c = 0.0342$ K/m is a constant, P_{atm} denotes the atmospheric pressure in Pa, h is the height or distance in m, T_{out} and T_{in} are the outdoor and indoor temperature respectively.

Using the average basement radon activity concentrations under stack open and closed conditions (mostly a successive 30 day period), the mean effectiveness/efficiency of a PSD system in achieving reduction of indoor radon levels termed as Percentage of radon reduction can be calculated (Zhou et al. 2021):

$$\text{Percentage of radon reduction} = \left(\frac{C_{Rn\ \text{stack closed}} - C_{Rn\ \text{stack open}}}{C_{Rn\ \text{stack closed}}}\right) \times 100\% \tag{8.2}$$

where $C_{Rn\ \text{stack open}}$ is the indoor radon activity concentration when the radon stack is open/operational (Bq/m^3) and $C_{Rn\ \text{stack closed}}$ denotes the indoor radon activity concentration when the radon stack is closed/non-operational (Bq/m^3).

(i) Conditions for PSD installation

The efficiency of a PSD system is dependent on factors like thermal buoyancy of air in the vent pipe and its ability to depressurize the soil underneath the building. The following conditions must be satisfied to set up an efficient PSD system (Zhou et al. 2021):

(i) A uniform permeable layer viz., concrete slabs, crawlspace membranes etc. directly in contact with the soil;
(ii) Proper insulation of the vent pipe; discharge from the exhaust duct above the highest roof;
(iii) Proper labeling of the systems for differentiating it with the plumbing system;
(iv) Proper routing of vent pipe for easy re-installation of fan to adjust the system when the performance gets poor in future;
(v) Proper sealing of the components in contact with the ground to avoid soil gas infiltration.
(vi) Provisions for continuous or periodic monitoring of the system performance in regular intervals.

A study conducted in Ireland on six unoccupied dwellings of identical construction under three controlled conditions, i.e., closed passive sump, open passive sump and open passive sump with a static cowl installed (Monahan et al. 2022). This radon testing were carried out over a six week period and the highest reduction rate of radon levels (~75%) were observed with the fitting of passive sump coupled with static cowl. In brief, passive sump with a static cowl in new dwellings was found to be highly efficient and cost effective (~€100 per dwelling) for bringing down radon levels below Irish reference levels for dwellings, i.e., 200 Bq/m^3 (Lacchia et al. 2020; Monahan et al. 2022).

b. Active soil depressurization (ASD)

The basic principle of active soil depressurization (ASD) technique is to reduce radon gas entry by reversing the air-pressure relationship between the indoors and the radon source. i.e., soil using an inline fan to create suction (McAlary et al. 2020). ASD is easier to install and far efficient in radon reduction compared to PSD. In this way, ASD is a reliable option for engineers and builders. The principal components of ASD are suction point in a continuous, homogeneous, permeable aggregate layer, groundwater control system, or a sump beneath the ground contacted floor of the dwelling (McAlary 2021).

The Fig. 8.2 illustrates an ASD system with its components. Similar to PSD, an ASD system should also include certain components, and the following conditions should be achieved for installation (Knapp et al. 2020; McAlary 2021):

(i) Suction point located beneath the soil or slab in contact with the floor;
(ii) A continuous and homogenous permeable aggregate layer, sump or groundwater control system;
(iii) Location of discharge exit point above the highest roof;
(iv) Continuously operating energy efficient inline fan;
(v) Proper labeling of systems for differentiating it with other plumbing systems;

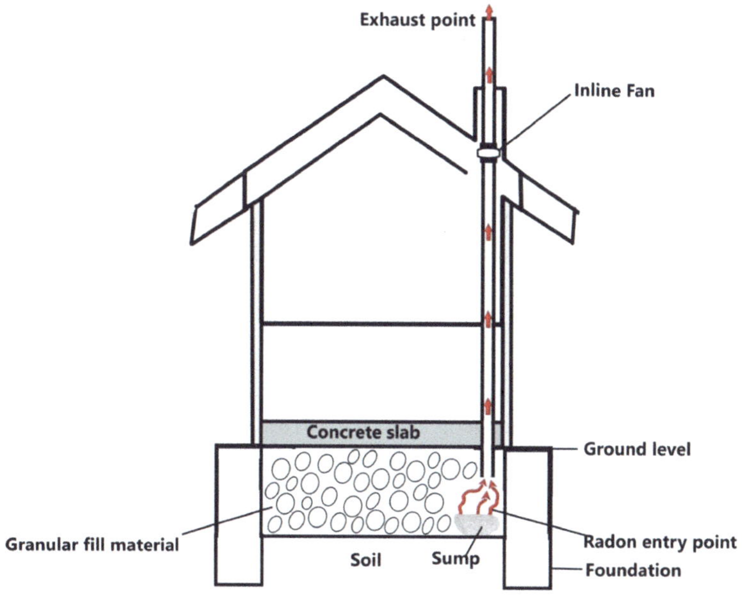

Fig. 8.2 Active soil depressurization for radon control in new constructions

(vi) U-tube manometer to check the performance regarding pressure differences in the vent pipe.

Sub-slab depressurization (SSD) system is a commonly utilized variant of the active soil depressurization (ASD) technique, designed to mitigate radon infiltration in buildings. Typically, one or more suction pipes are installed beneath a concrete slab, and the soil gas collected using suction is directed through these pipes and exhausted outdoors. To enhance the efficiency of SSD systems, a sump pumping well (SPW) framework installed in the soil adjacent to the wall can be employed as an energy-saving alternative during the soil air extraction process (Saâdi et al. 2021).

When a vapor barrier or membrane is installed below the floor or over the soil, and suction pits are placed below the membrane, the resulting system is called a sub-membrane depressurization system (Chen et al. 2015). These systems create negative pressure below the building, preventing radon gas from infiltrating the dwelling with lower pressure than the sub-surface. Another technique is drain-tile depressurization (DTD), which involves placing the suction point at a drain tile either inside or outside the footings of the dwelling (Shubin et al. 2021). Block-wall depressurization (BWD) is another approach that uses one or multiple pipes connected to the void network within a block foundation wall and a fan (Baskaran et al. 2016).

c. Sealing of surfaces

Sealing the surfaces that separate indoor occupied spaces from the ground can be a highly effective method when used in conjunction with PSD or ASD systems (IAEA 2021; Zhou et al. 2021). This practice can enhance the performance of soil depressurization systems by reducing conditioned air loss and increasing the air pressure reversal rate from soil stratum to indoor environment. The choice of sealant depends on the surface or foundation type. For instance, as the floors in crawlspaces are often open gravel or dirt, the most commonly used sealing method practiced is to seal the dirt floor using high density polyethylene (e.g., durable plastic, rubber, cross laminate etc.) sheeting (Lotz 2021). Polyurethane based caulk may be used as sealant for cracks in concrete slabs. While sealing of cracks and openings is considered as a basic recommended step for most other mitigation techniques, it is not a great option for reducing radon when used separately.

d. Barriers and membranes

Barriers or membranes placed between the indoor space and the soil provide an effective solution to this issue (Jeng and Zhai 2018). While this option can be used as a stand-alone method, it is often coupled with soil depressurization systems for improved performance. However, achieving airtightness of the membranes under typical construction conditions can be challenging, and extreme temperature differences in certain regions may limit the use of this approach (Felicioni et al. 2023). For an effective barrier, it is essential to consider properties such as durability, strength, and architectural compatibility, as well as sufficiently low gas permeability to minimize radon diffusion through the material (Cosma et al. 2015; Jiránek and

Kačmaříková 2019). These barriers can be installed as membranes or films, or as a flowable material applied to the construction matrix.

e. Ventilation of unoccupied spaces

Ventilating unoccupied spaces between the soil and occupied space, such as vented crawlspaces, is a viable option for reducing radon activity concentration. By separating the indoors from the soil stratum and lowering radon levels beneath the occupied space, this method can effectively minimize radon infiltration. However, the success of this method hinges on the air tightness of the floor above the vented unoccupied space (Hernandez et al. 2020). Passive ventilation, which considers vent distribution around the perimeter of the unoccupied space, is one approach.

In some cases, the unoccupied space is subjected to either pressurization or depressurization using a fan. However, this approach is a less desirable option as it could lead to the back-drafting of combustion appliances and energy loss (Rathgeber et al. 2019). Either active or passive type of sub slab and sub membrane depressurization may be used as a better alternative for crawlspace ventilation, as these two methods could facilitate greater rates of radon reduction in a building.

f. Ventilation of occupied spaces

As a general thumb rule, a considerable degree of air exchange between indoor and outdoor environments is favorable for achieving and maintaining superior indoor air quality in dwellings (Mareș et al. 2021). The effects of ventilation on indoor radon levels are highly variable, especially in regions with extreme climatic conditions. This method is best suitable for dwellings where the building material itself is the major source of radon. In practical terms, it is not easy for testing the radon levels in building materials prior to construction. That being the case, this approach could serve as a feasible solution for reducing the radon levels. Casado-Vara et al. (2021) proposed a new strategy using hybrid algorithm for automatic radon gas ventilation in smart buildings.

g. Water treatment

In regions where the radon activity concentrations in water are high and a matter of health concern, water treatment is a viable method for radon reduction. More details on various methods of water treatment are explained in Sect. 8.9.

8.6 Determinants of Efficiency in Soil Depressurization Systems

Recent studies have highlighted the importance of considering the permeability of the aggregate layer in the design of soil depressurization systems (Goggins et al.

2019; Hung et al. 2018; Fuente et al. 2019). To ensure the effectiveness of sub-slab depressurization systems, a thorough understanding of fluid physics is crucial (Frutos et al. 2020). This requires the use of robust numerical models and accurate assessment of radon activities in basements. In selecting a model for estimating the radon entry rate into the basement, it is essential to choose an appropriate model to ensure accurate results. Saâdi et al. (2021) found meshed-basement model to be more efficient than classical mass balance model in producing more accurate radon entry rate simulations.

8.7 Radon Mitigation Strategies in Existing Buildings

Indoor radon reduction measures can be broadly classified into two categories:

(i) Prevention of radon entry into dwellings by altering the indoor-outdoor pressure gradient and sealing the surface;
(ii) Removal of radon post-entry into dwellings using appropriate cost effective methods.

Various preventive measures applied for reducing indoor radon levels in new buildings were discussed in the previous sections. Radon mitigation is mostly based on the same principles followed by preventive strategies yet with some subtle variations. Measurement of indoor radon is the first step prior to selection of the mitigation strategy. Active soil depressurization is one of the most efficient mitigation measures reported by several studies. Nunes et al. (2022) addressed the need to develop mitigation systems with smart ventilation mechanisms.

Development of more cost-effective mitigation strategies will further promote their installation in areas of high radon potential. In order to resolve affordability issues, radon control incentives were implemented in Iowa (United States) and this attempt witnessed significant increase of radon testing and mitigation rates subsequently (Bain et al. 2016). Other than efficiency and affordability, the choice of mitigation techniques must be in consistent with the recognized protocols as well as recommended reference levels followed by that particular country/state/region.

8.8 Factors Affecting Radon Mitigation

The major factors to be considered to select the suitable mitigation approach for a building are: age of the building, basement geological data, initial indoor radon level, radon exhalation rate, meteorological data, soil data etc.

In majority of the countries, the pre-mitigation diagnosis is conducted by a private contractor assigned to install mitigation systems. Failure in carrying out a proper thorough examination of a building prior to mitigation results in selecting an inappropriate technique (Brossard et al. 2015). In this regard, the following two broad phases should be adhered during the diagnostic procedure:

(i) **Primary phase (Visual inspection of the dwelling)**: This phase is important to determine the applicable mitigation strategy based on the radon entry dynamics which include radon entry points, locations apt for ASD suction point installation, routing options for ASD ducts, possible major depressurization sources, history of the construction and modifications done over the years, locations suitable for combustion appliances through which radon is pushed outdoors etc.
(ii) **Secondary phase**: Understanding the pressure differences, pressure field extension in the soil beneath the occupied space etc. This can be inspected using a non-thermal smoke ampoule or a micro-manometer. The air-tightness of the building should be determined to pressurize indoor occupied spaces or to dilute radon gas after entering indoors.

Measurement of air flow rate is necessary for understanding the ventilation rate. In mechanically ventilated buildings, continuous radon monitoring is recommended to check the effectiveness of mechanical ventilation system. In cases where high radon emanation rate from building materials is suspected, measurement on this aspect needs to be carried out. Groundwater samples from the private well should be collected and analyzed based on the geogenic radon index in the region.

The continuous variation of meteorological conditions, lifestyle of the inhabitants (use of AC, furnace, exhaust fans, opening and closing of windows etc.) also influence the radon dynamics diurnally (Schubert et al. 2018). Other than the aforementioned conditions, another important factor affecting mitigation in the recent days is the energy efficiency alterations in buildings. Being a topical issue, this aspect is included as a separate discussion in the forthcoming section.

8.9 Energy Efficiency and Indoor Radon

Since the 1980s, a very common conviction has existed that energy-efficient dwellings tend to accumulate higher radon levels. In order to break this notion, Chi and Laquatra (1990) made a preliminary attempt to study the relationship between energy efficiency and indoor radon risks. Research in this domain has been carried out till date. A notable increase of radon activity concentration by 33% was noticed in high radon potential regions under the national energy saving program in the United States (Pigg et al. 2018). Energy retrofitting reduces ventilation, resulting in increased radon levels in such buildings.

Thermal retrofitting, one of the commonly used approaches to attain energy-efficiency in buildings by decreasing air infiltration and increasing thermal insulation of dwelling envelop has been associated with increased radon build up. There is a possibility of two to three-fold increase of radon activity concentration in energy retrofitted houses than conventional dwellings (Jiránek and Kačmaříková 2014; Meyer 2019). In a recent study, radon levels in 650 energy efficient dwellings were investigated in Switzerland (Yang et al. 2019). The results of the study suggested that energy renovation measures without taking account of indoor environment could have adverse impacts on indoor radon levels.

Yarmoshenko et al. (2020) developed an index viz., Energy Efficiency Index (EEI) to quantify the energy efficiency potential of a dwelling in terms of reducing uncontrolled ventilation to analyze the relationship between degrees of energy efficiency and measured indoor radon activity concentrations. After assigning EEI for each building, it was concluded that the mean radon activity concentrations in new age multi-storey apartment energy-efficient buildings are higher compared to typical flat houses. Radon index, i.e., radon potential based on geological basement is considered for identifying radon hotspot areas. Nevertheless, it is necessary to apply radon preventive measures in newly constructed dwellings, irrespective of the radon potential. In this regard, long-term passive indoor measurements are recommended for assessing radon risk (Florică et al. 2020).

Burghele et al. (2021) selected individual energy efficient dwellings ($n = 30$) using different mitigation systems with a real time indoor monitoring at high radon potential sites in Romania. The best performance in radon reduction was recorded in houses with sub-slab depressurization (61–95%) followed by underground sump depressurization (73–85%).

The efficiency of each radon reduction system could be assessed based on radon activity concentrations measured using passive detectors, before (E1) and after (E2) intervention in various rooms of the building and reported as house average, using the following equation (Burghele et al. 2021):

$$\text{Efficiency}\ (\%) = \frac{E1 - E2}{E2} \times 100\% \tag{8.3}$$

8.10 Radon Removal Strategies in Groundwater

Certain techniques for reducing or eliminating radon in groundwater have been developed. A few of such strategies are very efficient in removing high radon levels in groundwater system before it enters the building.

8.10.1 Heating Techniques

Heating is one of the simple techniques available to remove radon in potable water (Ghernaout 2019). In this technique, heating starts from the surface of material which often leads to uneven distribution of heat. Also, this results in long heating times and high power consumption. These drawbacks can be overcome by using microwave technique instead of conventional heating method (Grekov et al. 2020). As microwave penetrates into the object with the speed of light, uniform heating of the surface can be achieved. In this way, heating time and energy use can be efficiently improved. But this method can increase radon activity in air, which is another demerit (Ghernaout 2019).

8.10.2 Membrane/Filtration Techniques

Reverse osmosis (RO) systems reduce the radon activity concentrations using semi-permeable membranes (0.0001–0.001 μm) and pressure to separate dissolved species from water. Radon removal in the range of 78.7–96.5% has been reported using reverse osmosis or electrodialysis processes along with aeration (Alabdula'aly 1999). However, this technique has some disadvantages as well. In the first place, RO systems are expensive. Secondly, the membranes have poor functioning abilities when interrupted by a precipitate (e.g., Fe, Mn, Si, Ca etc.) in the water (NRC 1999). Biological activated carbon membrane bioreactors (BAC-MBR) seem to be a better option (Hasan and Muhammad 2020).

8.10.3 Aeration Techniques

Aeration is one of the primary techniques applied for removing radon in water (Nayak et al. 2022). The efficiency of this method depends on pH enhancement due to the carbon dioxide degassing during the process. It can be carried out in two ways—either releasing water into air or releasing air into water. However, there is a need to design more cost-effective aeration systems for using it on a private basis (Sahoo et al. 2021). Several variants of this technique are practiced (e.g., packed-tower, spray, diffused bubble, tray, cascade, pressure aeration etc.) depending on the requirement level (NRC 1999). One of the most preferred methods is the packed-tower aeration technique where water to be treated is sprayed into the tower top and trickled down over a plastic packing with high surface area to volume ratio. Proper maintenance and monitoring of aeration system is required because of its proneness to technical failures during operation (Pradhan and Sukla 2019).

8.10.4 Granular Activated Carbon (GAC)

Granular activated carbon (GAC) is made of materials viz., wood, bone, coconut shells etc. subjected to high heat and pressure in order to activate them by increasing the surface area. This improves the adsorption capacity and GAC is suitable for adsorbing organic chemicals and dissolved gases (NRC 1999). Being an inert gas, radon could be effectively adsorbed and removed using GAC and this method is especially suitable for pilot level studies (Fonollosa et al. 2015).

8.10.5 *Biological Techniques*

Research on microbes and microalgae prove that the activated intercellular polysaccharides of biological cell of these organisms mimic the adsorptive properties of activated carbon (Nayak et al. 2022). This property enables them to remove radon in the same way a GAC approach does. Unlike the radio-vulnerable higher organisms, microbes are radio-resistant due to the extremozymes and extremolytes present in them (Nayak et al. 2021a, b). For example, members of the genus *Kocuria* has been isolated from radon-rich underground spring waters (Timkina et al. 2022).

Microbial elimination of radon involves the remediation via precipitation, bioadsorption, biotransformation and bioaccumulation processes (Pradhan and Sukla 2019). The metabolic activity of microbes facilitates the conversion of soluble or toxic radioactive compounds to insoluble less toxic or non-radioactive forms. Radio-resistant microbes and activated carbon could be used separately or combined (biologically activated carbon) for a better performance in radon elimination. This is a cost-effective method if appropriate care is provided during the whole process. Membrane bioreactor (MBR) method is a hybrid technique (combination of biological as well as membrane techniques) proven to be effective in removing radionuclides in water (Hasan and Muhammad 2020).

From the literature it is evident that aeration, activated carbon and biological are the commonly preferred and feasible methods in the present day. However, analyzing the composition of the source water is important to avoid clogging or fouling inside the treatment systems.

8.11 Conclusion and Recommendations

Many countries have implemented various measures to mitigate radon, but the lack of systematic follow-up measurements creates uncertainty about the long-term effectiveness of these measures. Active soil depressurization has been shown to be the most effective technique for reducing radon levels, but it can be expensive and disruptive to install in occupied homes. To ensure sustained low levels of radon, proper maintenance and the installation of alarms that detect motor failures or electrical outages (power failures) are necessary.

The impact of new technologies used in energy-efficient building construction on radiation doses should be examined in the context of radiation protection. Therefore, it is necessary to promote education and awareness among building engineers and the general public, establish public funding for radon prevention and mitigation, identify priority areas for radon monitoring and prevention in land use maps, and incorporate relevant regulations and guidelines into building codes. To achieve these objectives, continuous communication among scientists, professionals, the public, and stakeholders is essential. This calls for multidisciplinary research in physics, radiation protection, geology, construction engineering, and materials science.

References

Akbari K, Mahmoudi J, Ghanbari M (2013) Influence of indoor air conditions on radon concentration in a detached house. J Environ Radioact 116:166–173

Alabdula'aly AI (1999) Occurrence of radon in the central region groundwater of Saudi Arabia. J Environ Radioact 44(1):85–95

Alabid J, Bennadji A, Seddiki M (2022) A review on the energy retrofit policies and improvements of the UK existing buildings, challenges and benefits. Renew Sustain Energy Rev 159:112161

Bain AA, Abbott AL, Miller LL (2016) Successes and challenges in implementation of radon control activities in Iowa, 2010–2015. Prev Chronic Dis 13:150596

Bastian Z, Schnieders J, Conner W, Kaufmann B, Lepp L, Norwood Z, Simmonds A, Theoboldt I (2022) Retrofit with passive house components. Energ Effi 15(1):10

Brossard M, Falcomer R, Whyte J (2015) Radon mitigation in cold climates at Kitigan Zibi Anishinabeg. Health Phys 108(1):S13–S18

Burghele BD, Botoş M, Beldean-Galea S, Cucoş A, Catalina T, Dicu T, Dobrei G, Florică Ş, Istrate A, Lupulescu A, Moldovan M (2021) Comprehensive survey on radon mitigation and indoor air quality in energy efficient buildings from Romania. Sci Total Environ 751:141858

Casado-Vara R, García-Retuerta D, Bartolomé Á, Jove E, Calvo-Rolle JL, Rey AMD, Corchado JM (2021) Demand control ventilation strategy by tracing the radon concentration in smart buildings. In: 15th international conference on soft computing models in industrial and environmental applications (SOCO 2020), vol 15. Springer International Publishing, pp 374–382

Chen J, Whyte J, Ford K (2015) An overview of radon research in Canada. Radiat Prot Dosimetry 167(1–3):44–48

Chi PS, Laquatra J (1990) Energy efficiency and radon risks in residential housing. Energy 15(2):81–89

Cori L, Curzio O, Donzelli G, Bustaffa E, Bianchi F (2022) A systematic review of radon risk perception, awareness, and knowledge: risk communication options. Sustainability 14(17):10505

Cosma C, Papp B, Cucoş A, Sainz C (2015) Testing radon mitigation techniques in a pilot house from Băiţa-Ştei radon prone area (Romania). J Environ Radioact 140:141–147

Felicioni L, Jiránek M, Lupíšek A (2023) Environmental impacts of waterproof membranes with respect to their radon resistance. Sustain Mater Technol 35:e00541

Finne IE, Kolstad T, Larsson M, Olsen B, Prendergast J, Rudjord AL (2019) Significant reduction in indoor radon in newly built houses. J Environ Radioact 196:259–263

Florică Ş, Burghele BD, Bican-Brişan N, Begy R, Codrea V, Cucoş A, Catalina T, Dicu T, Dobrei G, Istrate A, Lupulescu A (2020) The path from geology to indoor radon. Environ Geochem Health 42(9):2655–2665

Fonollosa E, Nieto A, Peñalver A, Aguilar C, Borrull F (2015) Presence of radionuclides in sludge from conventional drinking water treatment plants. A review. J Environ Radioact 141:24–31

Frutos B, Martín-Consuegra F, Alonso C, Perez G, Peón J, Ruano-Ravina A, Barros JM, Santorun AM (2021) Inner wall filler as a singular and significant source of indoor radon pollution in heritage buildings: an exhalation method-based approach. Build Environ 201:108005

Frutos B, Sicilia I, Campo O, Aparicio S, González M, Anaya JJ, Rábago D, Sainz C (2020) A full-scale experimental study of sub-slab pressure fields induced by underground perforated pipes as a soil depressurisation technique in radon mitigation. J Environ Radioact 225:106420

Fuente M, Rábago D, Goggins J, Fuente I, Sainz C, Foley M (2019) Radon mitigation by soil depressurisation case study: radon concentration and pressure field extension monitoring in a pilot house in Spain. Sci Total Environ 695:133746

Gaskin J, Li YE, Ganapathy G, Nong G, Whyte J, Zhou LG (2021) Radon infiltration building envelope test system: evaluation of barrier materials. Radiat Prot Dosimetry 196(1–2):17–25

George AC (2015) The history, development and the present status of the radon measurement programme in the United States of America. Radiat Prot Dosimetry 167(1–3):8–14

Ghernaout D (2019) Aeration process for removing radon from drinking water—a review. Appl Eng 3(1):32–45

Goggins J, Croxford C, Foley M (2019) Large-scale experimental investigations of specified granular fill materials for radon mitigation by active and passive soil depressurisations. J Environ Radioact 207:27–36

Grekov D, Pré P, Alappat BJ (2020) Microwave mode of heating in the preparation of porous carbon materials for adsorption and energy storage applications—An overview. Renew Sustain Energy Rev 124:109743

Gruber V, Bossew P, De Cort M, Tollefsen T (2013) The European map of the geogenic radon potential. J Radiol Prot 33(1):51

Gulan L, Stajic JM, Spasic D, Forkapic S (2022) Radon levels and indoor air quality after application of thermal retrofit measures—a case study. Air Qual Atmos Health 1–11

Hasan HA, Muhammad MH (2020) A review of biological drinking water treatment technologies for contaminants removal from polluted water resources. J Water Process Eng 33:101035

Hernandez G, Wallis SL, Graves I, Narain S, Birchmore R, Berry TA (2020) The effect of ventilation on volatile organic compounds produced by new furnishings in residential buildings. Atmos Environ X 6:100069

Hodgson SA, Pudner V (2019a) Passive remediation for radon in UK homes. Centre for Radiation, Chemical and Environmental Hazards, Public Health England

Hodgson SA, Pudner V (2019b) Passive remediation for radon in UK homes. Report PHE-CRCE-045. Public Health England, Oxfordshire, United Kingdom

IAEA (International Atomic Energy Agency) (2021) Protection against exposure due to radon indoors and gamma radiation from construction materials—methods of prevention and mitigation. IAEA-TECDOC1951, 2021. Accessed 12 Nov 2022

Ivanova K, Stojanovska Z, Tsenova M, Kunovska B (2017) Building-specific factors affecting indoor radon concentration variations in different regions in Bulgaria. Air Qual Atmos Health 10:1151–1161

Jeng CY, Zhai X (2018) Vapor intrusion. In: Fundamentals of environmental site assessment and remediation. CRC Press, pp 143–173

Jiránek M, Kačmaříková V (2014) Dealing with the increased radon concentration in thermally retrofitted buildings. Radiat Prot Dosimetry 160(1–3):43–47

Jiránek M, Kačmaříková V (2019) Radon diffusion coefficients and radon resistances of waterproofing materials available on the building market. J Environ Radioact 208:106019

Khan SM, Gomes J, Krewski DR (2019) Radon interventions around the globe: a systematic review. Heliyon 5(5):e01737

Khan SM, Pearson DD, Rönnqvist T, Nielsen ME, Taron JM, Goodarzi AA (2021) Rising Canadian and falling Swedish radon gas exposure as a consequence of 20th to 21st century residential build practices. Sci Rep 11(1):17551

Knapp S, Athey JE, Nash A (2020) Mitigating radon levels at home. State of Alaska, Department of Natural Resources, Division of Geological & Geophysical Surveys

Lacchia AR, Schuitema G, Banerjee A (2020) "Following the science": in search of evidence-based policy for indoor air pollution from radon in Ireland. Sustainability 12(21):9197

Lotz WA (2021) Moisture control and insulation systems in buildings, chilled water pipes and underground pipes: a guide for architects, engineers, contractors, facility managers, construction professionals and homeowners. Universal-Publishers

Mareş IC, Catalina T, Istrate MA, Cucoş A, Dicu T, Burghele BD, Hening K, Popescu LL, Popescu RS (2021) Research on best solution for improving indoor air quality and reducing energy consumption in a high-risk radon dwelling from Romania. Int J Environ Res Public Health 18(23):12482

Mc Laughlin JP, Gutierrez-Villanueva JL, Perko T (2022) Suggestions for improvements in national radon control strategies of member states which were developed as a requirement of EU directive 2013/59 EURATOM. Int J Environ Res Public Health 19(7):3805

McAlary T, Wertz W, Mali D, Nicholson P (2020) Mathematical analysis and flux-based radius of influence for radon/VOC vapor intrusion mitigation systems. Sci Total Environ 740:139988

McAlary TA (2021) Subslab depressurization versus subslab ventilation: insights from recent research. Groundw Monit Remed 41(2):132–137

Meyer W (2019) Impact of constructional energy-saving measures on radon levels indoors. Indoor Air 29(4):680–685

Monahan E, Murphy P, Long S, Dowdall A (2022) The effectiveness of passive sumps and static cowls in reducing radon levels in new build Irish dwellings. J Environ Radioact 248:106866

National Research Council (1999) Risk assessment of radon in drinking water. National Academies Press

Nayak T, Basak S, Deb A, Dhal PK (2022) A systematic review on groundwater radon distribution with human health consequences and probable mitigation strategy. J Environ Radioact 247:106852

Nayak T, De D, Karmakar P, Deb A, Dhal PK (2021a) Microbial communities of the drinking water with gradient radon concentration are primarily contributed by radon and heavy metal content. Front Environ Sci 156

Nayak T, Sengupta I, Dhal PK (2021b) A new era of radiation resistance bacteria in bioremediation and production of bioactive compounds with therapeutic potential and other aspects: an in-perspective review. J Environ Radioact 237:106696

Norwegian Ministries (2010) Strategy for the reduction of radon exposure in Norway. Norwegian Ministry of Labour, Norwegian Ministry of Health and Care Services and Norwegian Ministry of Local Government and Regional Development. Norwegian Government Administration Services Publication, Oslo

NRPA (2013) Measurement protocol for radon in dwellings (in Norwegian). http://www.nrpa.no/filer/e33cd9ffab.pdf. Accessed 05th Mar 2023

Nunes LJ, Curado A, Graça LCD, Soares S, Lopes SI (2022) Impacts of indoor radon on health: a comprehensive review on causes, assessment and remediation strategies. Int J Environ Res Public Health 19(7):3929

Pigg S, Cautley D, Francisco PW (2018) Impacts of weatherization on indoor air quality: a field study of 514 homes. Indoor Air 28(2):307–317

Pradhan D, Sukla LB (2019) Removal of radon from radionuclide-contaminated water using microalgae. In: The role of microalgae in wastewater treatment. Springer, Singapore, pp 75–86

Rathgeber W, Parimi M, Norouzalizadeh R, Lee I, Liu H (2019) A critical recommendation to The University of British Colombia's green building action plan, for: implementing standardised human factor policy and metrics in the emergent component of health and well-being for institutional buildings. https://doi.org/10.14288/1.0386716

Ruvira B, García-Fayos B, Juste B, Arnal JM, Verdú G (2022) Experimental estimation of the diffusion coefficient in radon barrier materials based on ISO/TS 11665–13: 2017. Radiat Phys Chem 193:109993

Saâdi Z, Mansouri N, Marie L, Ielsch G (2021) Towards a better assessment model of transient radon concentrations in dwellings basements for the study of the effectiveness of soil radon mitigation systems designs. J Environ Radioact 235:106651

Sahoo T, Tripathy R, Panda J, Hembram M, Soren S, Rath CK, Sahoo SK, Sahu R (2021) Radionuclides: availability, effect, and removal techniques. In: Applied water science, vol 1: fundamentals and applications, p 331

Schubert M, Musolff A, Weiss H (2018) Influences of meteorological parameters on indoor radon concentrations (222Rn) excluding the effects of forced ventilation and radon exhalation from soil and building materials. J Environ Radioact 192:81–85
Shubin IL, Bakaeva NV, Kalaydo AV (2021) Comparison of approaches to ensuring the radon safety of designed buildings in Russia and USA. In: IOP conference series: materials science and engineering, vol 1030, No. 1. IOP Publishing, p 012027
Silva AS, Dinis ML (2021) An overview of the development and implementation of the radon action plans in European countries. In: Arezes PM, Baptista JS, Barroso MP, Carneiro P, Cordeiro P, Melo RB, Miguel AS, Perestrelo G (eds) Occupational and environmental safety and health. CRC Press/Balkema, pp 225–235
Stanley FK, Irvine JL, Jacques WR, Salgia SR, Innes DG, Winquist BD, Torr D, Brenner DR, Goodarzi AA (2019) Radon exposure is rising steadily within the modern North American residential environment, and is increasingly uniform across seasons. Sci Rep 9(1):18472
Stanley FK, Zarezadeh S, Dumais CD, Dumais K, MacQueen R, Clement F, Goodarzi AA (2017) Comprehensive survey of household radon gas levels and risk factors in southern Alberta. Canadian Med Assoc Open Access J 5(1):E255–E264
Timkina E, Drábová L, Palyzová A, Řezanka T, Maťátková O, Kolouchová I (2022) Kocuria strains from unique radon spring water from Jachymov Spa. Fermentation 8(1):35
Vogeltanz-Holm N, Schwartz GG (2018) Radon and lung cancer: what does the public really know? J Environ Radioact 192:26–31
Wang B, Song T, Li J, Zhang X, Li Y, Han X, Yang X (2022) Design on intermittent ventilation strategy for radon removal in underground space. J Environ Radioact 253:107007
World Health Organization (2007) International radon project: survey on radon guidelines, programmes and activities (No. WHO/HSE/RAD/07.01). World Health Organization
Yang S, Goyette Pernot J, Hager Jörin C, Niculita-Hirzel H, Perret V, Licina D (2019) Radon investigation in 650 energy efficient dwellings in western Switzerland: impact of energy renovation and building characteristics. Atmosphere 10(12):777
Yarmoshenko IV, Onishchenko AD, Malinovsky GP, Vasilyev AV, Nazarov EI, Zhukovsky MV (2020) Radon concentration in conventional and new energy efficient multi-storey apartment houses: results of survey in four Russian cities. Sci Rep 10(1):1–14
Zhou LG, Berquist J, Li YE, Whyte J, Gaskin J, Vuotari M, Nong G (2021) Passive soil depressurization in Canadian homes for radon control. Build Environ 188:107487

SPRINGER NATURE

GPSR Compliance

The European Union's (EU) General Product Safety Regulation (GPSR) is a set of rules that requires consumer products to be safe and our obligations to ensure this.

If you have any concerns about our products, you can contact us on ProductSafety@springernature.com

In case Publisher is established outside the EU, the EU authorized representative is:

Springer Nature Customer Service Center GmbH
Europaplatz 3
69115 Heidelberg, Germany

The manufacturer's authorised representative in the EU is Springer Nature Customer Service Centre GmbH, Europaplatz 3, 69115 Heidelberg, Germany. If you have any concerns regarding our products, please contact ProductSafety@springernature.com

Printed and bound by CPI Group (UK) Ltd, Croydon, CR0 4YY
01/10/2026
02225893-0003